Short Sunderland

Short Sunderland

Ken Delve

The Crowood Press

First published in 2000 by
The Crowood Press Ltd
Ramsbury, Marlborough
Wiltshire SN8 2HR

British Library Cataloguing-in-Publication Data
A catalogue record for this book is available from the British Library.

ISBN 1 86126 355 4

Photograph previous page: a Sunderland V with an ASV blister under the wing. Ken Delve Collection.

Typefaces used: Goudy (*text*),
Cheltenham (*headings*).

Typeset and designed by
D & N Publishing
Membury Business Park, Lambourn Woodlands
Hungerford, Berkshire.

Printed and bound in Great Britain by
Bookcraft, Bath.

Acknowledgements

Much of this book has been compiled from official records and I am indebted to the staff of the Air Historical Branch for their help in this regard. Sadly, there is a dwindling number of Sunderland survivors as we reach more than fifty years from the end of the Second World War – but I am very grateful to those survivors who have helped with this book, especially to Dr Arthur Banks for his help with contacts and photographs. Thanks go to my two aviation history colleagues Andy Thomas and Peter Green for their help with photographs.

Contents

Introduction

'On the step'; classic shot of a Short Sunderland flying boat. These aircraft played an important role in the maritime war, flying thousands of hours of cover for Allied convoys. Ken Delve Collection

> The Sunderland has a very wide range and an armament formidable enough for it to be nicknamed the 'flying porcupine' (*fliegende Stachelschwein*) by the Germans. It is the largest aircraft in use by Coastal Command and can almost be described as commodious. Meals can be cooked on board and there are bunks where those of the crew not on watch can sleep very comfortably. So much so that once a Sunderland dropped depth charges on a U-boat, and the second pilot, at rest and asleep at the time, knew nothing about the attack until he came on watch again an hour or so later. (*Coastal Command*, HMSO)
>
> The rugged aspect of the Sunderland may perhaps mislead the landsman who can be excused for thinking that such a craft should surely be quite at home in the open sea. But for all its ark-like proportions the Sunderland and flying boats generally are mere eggshells of marine architecture. They are essentially aircraft which float and not boats which fly.

Experienced Sunderland pilot and squadron CO Dundas Bednal has recorded many interesting snippets in his excellent book *Sun on my Wings*:

> The Sunderland formed a self-contained fighting unit which could maintain itself for long periods. It was fully equipped with bunks for the crew and could, therefore, be self supporting away from base. The only essentials in this case would be access to fuel and food supplies and ammunition and weapon replacements.
>
> Each aircraft, therefore, had its own 'team' which led to a high standard of serviceability. When flying, the fitters and riggers became air gunners and whilst approaching a mooring after landing they became seamen performing the mooring operation.

The Sunderland's main wartime role was anti-submarine warfare, in which it scored around thirty submarine 'kills' (figures vary), and many other sightings and attacks that either caused damage or persuaded the enemy that the area was too dangerous. In addition, Sunderlands undertook a number of dramatic rescues, landing in hazardous conditions. The Sunderland story continued post-war, especially with the squadrons of the Far East Flying Boat Wing and operations over Malaya and Korea. Although a few aircraft have been preserved in various parts of the world, there is now only one airworthy Sunderland, based in Florida, and the aircraft rarely flies.

CHAPTER ONE

Pre-War Years

Shorts and the Flying Boat

Shorts entered the 'aircraft on water' market during World War One. A number of Shorts' designs entered service, the Short 184 single-engined biplane seaplane proving to be the most successful. Two prototypes, S106 and S107, were built in 1915 and were followed by substantial military orders, serving primarily with the RAF and Royal Naval Air Service. Indeed, the company's Rochester works received a large number of contracts from the Admiralty and such was the demand for the seaplane designs coming from Shorts, that a number of other companies licence-built the aircraft. Involvement with true flying boats, however, commenced in 1918, the company receiving a contract to build fifty F.3 and F.5 boats designed by John Porte at Felixstowe. The first F.3 built by Shorts, N4000, was hoisted into the water in May 1918 and production of the type continued into 1920, with a batch of F.5s, the first ones being completed for the Japanese Navy, completed in the middle of that year.

Oswald Short was impressed with the potential of such aircraft and in 1918 he submitted a tender to the Admiralty for specification Admiralty N3, for a twin-engined flying boat designed to work with the British Grand Fleet. In the immediate post-war period the requirement for new aircraft was limited, and the original order for three prototypes was reduced to one, this first flying on 19 April 1921 as the Short N3 Cromarty; its design was based on the Porte aircraft. This aircraft, N120, underwent a series of modifications before being handed to the Seaplane Development Flight in July the following year. In August the aircraft was damaged beyond repair in a taxying accident, ending any prospects it had of a military career.

This first venture into flying boats was not an immediate success and the basis of the company's business remained seaplanes, although F.5s continued to be built. However, the Air Ministry had ordered a single F.5 with duralumin hull and when this proved successful a new all-metal flying boat, similar in size to the Cromarty, was ordered: the Short Singapore I. Under

Shorts entered the 'aircraft on water' market with the Short 184 seaplane, the prototypes of which first flew in 1915. Ken Delve Collection

September 1937, Singapore III of 209 Squadron. Ken Delve Collection

Specification 13/24 a single prototype was ordered in 1925.

The first flight of the sole Singapore I, N179, took place off the Medway on 17 August 1926, powered by two Rolls-Royce Condors. Successful proving trials, including a Baltic 'cruise', showed the potential of the new flying boat. Alan Cobham then flew a 23,000-mile (37,000km) survey flight, on behalf of Imperial Airways, in the aircraft. The Singapore I was an important step in the development of Shorts' flying boats.

Imperial Airways ordered two examples of the three-engined Short Calcutta in 1927, these becoming the first commercial flying boats to enter airline service, opening an era of British commercial flying boat operation which culminated some twenty years later with the final cruises of the 'Empire' boats. The first Calcutta, G-EBVG, was launched in February 1928, making its first flight the following day. The first two Calcuttas entered service in October 1928, flying a route between Southampton and Guernsey. A few additional Calcuttas were bought by Imperial Airways (three) and the French Government (two), plus four licence-built for the French Navy. Whilst the Singapore I had not been taken up by the RAF for squadron service, they did adopt the Calcutta, naming it the Rangoon; six of these served with 203 Squadron at Basra in the Persian Gulf in the early 1930s.

Although the Condor engines of the Singapore I had been the death-knell of this aircraft, the Singapore II, powered by four Kestrel engines (two tractor and two pusher), was to prove more successful. The basic hull shape was the same as that of the Calcutta but significant improvements were made in respect of engine mountings and other drag-reducing measures. The prototype Singapore II, N246, was launched on 27 March 1930, and the type was selected by the RAF to replace its Rangoons and Supermarine Southamptons with overseas squadrons. The production version was the Singapore III and this entered RAF service with 210 Squadron at Pembroke Dock in January 1935, the first overseas squadron being 230 Squadron at Alexandria the following October.

A 1930 three-engined boat for Japan was followed by a four-engined version, the Kent, for Imperial Airways, only three of which were built. The last large biplane flying boat to be produced by the Rochester factory was the six-engined Sarafand, the prototype of which, S763, made its first flight on 9 July 1931. This aircraft was designed to have a transatlantic capability, streamlining being a major element of the design, but in many ways it was a logical growth of the Singapore II. The one and only Sarafand was test-flown by MAEE (Marine Aircraft Experimental Establishment) for about a year from mid-1932.

Shorts had gained a great deal of experience with, and a fine reputation for, its various flying boats, but production numbers had been small and in the inter-war period there were few military contracts, whilst the development of land-based airliners was making civil contracts hard to attract. The future was by no means guaranteed.

The 'Knuckleduster' was an important development stage for Shorts in their flying boat programme; K3574 was evaluated by 209 Squadron. Ken Delve Collection

First of Shorts' monoplane boats was the experimental, gull-winged Short 'Knuckleduster'. This interesting little aircraft was an invaluable stepping stone in terms of aerodynamic and performance data from the somewhat dated biplanes to the subsequent impressive series of monoplane 'Empire' boats. In 1933 Short Brothers had been asked to tender for a flying boat suitable for short-range and transatlantic operations, an aircraft with the same overall performance as the Sarafand but not the same bulk. It was considered that such an aircraft was essential for Imperial Airways development in the late 1930s. Initial design work produced an aircraft that showed great promise, Imperial Airways increasing their order from fourteen to twenty-eight aircraft. The Air Ministry also ordered a single prototype, under Specification R2/33, which in due course became the Sunderland.

Oswald Short and test pilot John Parker had looked at various design options, with particular attention being paid to the monoplane configuration, for which the Short Scion land-plane provided useful data. The height of the hull was made larger than strictly necessary in order to provide a simpler wing attachment scheme. The wing spar was a braced structure employing four T-sections, trussed vertically by tubular struts and horizontally by lattice compression beams. Low drag was a prime consideration and careful attention was paid to the design of the planing bottom; furthermore, all joints in the Alclad skinning were joggled and all rivets flush countersunk.

Of the first two Empire boats, G-ADHL 'Canopus' and G-ADHM 'Caledonia', the latter was equipped with additional fuel tanks (2,320gal/10,556ltr compared with 650gal/2956ltr) for long-range work over the Atlantic. 'Canopus' was launched on 2 July 1936, being handed over to Imperial Airways, after various trials and minor modifications, that October. Other Empire boats entered service with the airline over the following months, flying passenger routes but also playing a vital part in the Empire Air Mail Scheme. More details of civilian variants and use are included in a brief overview in Chapter 8.

The Sunderland is Born

Having been reasonably impressed with the Sarafand, the Air Ministry had, by mid-1933, decided that it needed a long-range flying boat as part of its general re-equipment programme. Tenders were sought under Specification R2/33 for a four-engined aircraft smaller than the Sarafand but with equivalent performance. As related above, the requirement coincided with the Imperial Airways requirement and Shorts were able to amalgamate the basic design parameters, the most significant decision taken by Oswald Short and his team being the choice of a monoplane design, despite the fact that all their previous large flying boats had been biplanes.

Although the Empire boats and the Sunderland were similar, they were not identical, the latter having various features required to meet the RAF's requirements for protective treatment, interchangeability, maintainability and armament. Tendering for this Specification, Shorts, under the designation S.25, built prototype K4774 whilst Saunders-Roe produced Saro A.33 K4773.

Both designs progressed well and in March 1936 the Air Ministry ordered a further eleven of each type, the Shorts contract being awarded under Specification 22/36. In summer 1936 the final design conference was held between Shorts and the Air Ministry, the discussion being largely based upon results from the early flights of 'Canopus' and the design considerations for the military variant. The only significant discussions concerned armament, a proposal for a forward-firing 37mm Vickers gun being deleted in favour of an FN11 turret, whilst the single 0.303in Lewis in the tail was replaced by a four-gun FN13 turret. These changes led to a significant design change; due to the alteration of the centre of gravity the wings had to be given a 4½-degree sweep back, and there was a corresponding rearward move of the main hull step. Prototype K4774 was flown in its original guise, the straight wing and 'in-line' engine arrangement making it significantly different in appearance to production Sunderlands. Taxying trials were conducted on 14 October 1937, the aircraft being powered by 950hp Bristol Pegasus X engines. Two days later the aircraft made its first flight, the crew comprising John Parker (pilot), Harold Piper (co-pilot) and George Cotton (engineer). Four flights were made over the next week or so, there being few problems other than the already-known tail-heaviness. The aircraft then went back into No. 3 Shop for the wing sweep and planing bottom/step changes. It was also given 1,010hp Pegasus XXII engines, re-emerging the following spring to resume flight trials. Light-load tests were followed by full-load tests (on 15–17 March) and handling and fuel consumption tests (29 March–8 April). The prototype was then flown to Felixstowe for evaluation by the MAEE.

The first production aircraft, L2158, made its maiden flight on 21 April 1938. This aircraft was retained by Shorts for further development work, initially in respect of gun turrets and autopilot. The second production aircraft, L2159, was flown to Felixstowe in early May to join the prototype for evaluation. However, on 9 June L2159 departed for the Far East for evaluation by 230 Squadron, its place at MAEE being taken by L2158.

Pre-War

The 210 Squadron historian recorded the changeover to the new aircraft:

> May 1938 was a significant month in the Squadron history because it brought the start of a new re-equipment programme and a change from biplane to monoplane aircraft in the shape of the Sunderland flying boat. The first of these aircraft had been flown to the Marine Aircraft Experimental Establishment at Felixstowe where four pilots were converted. The Squadron Commander, Wing Commander Plenderlieth, flew over from Felixstowe to Pembroke Dock in Sunderland L2159 in time for the Empire Air Day Display on 28 May. The Squadron received its own first two Sunderland aircraft on 24 June and during June and July selected crews also ferried Sunderland aircraft to Singapore for 230 Squadron based at Seletar (a total of eight had been ferried by early October). This ferrying continued at intervals throughout the year and one aircraft which left in December assisted Imperial Airways, by carrying as far as Penang a full load of mail destined for Australia as part of the GPO experiment to send all Empire mail by air without surcharge.

Imperial Airways Empire flying boat 'Calpurnia' (G-AETW); this aircraft crashed at Lake Habbaniya on 27 November 1938. Ken Delve Collection

Prototype Sunderland K4774 made its first flight on 16 October 1937, powered by four Bristol Pegasus Xs. Ken Delve Collection.

Sunderland I

Sunderland prototype K4774 first flew on 16 October 1937 and the first production Sunderland I (of an initial batch of eleven aircraft), L2158, flew on 21 April 1938. Final production of Sunderland Is ran to ninety aircraft. With its 112ft 9in wingspan and 85ft 4in length, the Sunderland I was a big aircraft; the basic dimensions of the Sunderland remained the same throughout its development history.

Wings: Two-spar all-metal structure cantilever monoplane.

Hull: Channel-section frames interconnected with 'Z-section' stiffeners and riveted sheeting.

Tail: Metal-covered cantilever monoplane unit with fabric-covered control surfaces.

Engines: Four 890hp (2,600rpm) Bristol Pegasus 22 nine-cylinder air-cooled radials, driving DH three-bladed two-pitch metal propellers.

Interior: Hull divided into two decks. Upper deck – forward, cockpit with side-by-side seating for two pilots, radio operator, navigator, and engineer. Lower deck – Frazer-Nash nose turret with two 0.303in guns, bomb-aimer's position, turret retracts for mooring procedure; mooring compartment is behind the turret, from where stairs lead to the upper deck. To right of stairs is a lavatory whilst a gangway to left of stairs leads to the officers' wardroom. Further aft is the galley, crew quarters and bomb 'compartment', along with pillar mountings for Vickers 0.303in guns; to the rear of the aircraft is a work bench, dinghy and various other equipment such as flame-floats and flares. At extreme end of the aircraft is the rear turret with four 0.303in guns.

MAP Specification dated 25 August 1942

Powerplant: Four Pegasus 22

Fuel/range: 1,720gal with maximum bombs, range 1,850 miles (235 miles less if Leigh Light fitted); 2,034gal with permanent tanks full, range 2,425 miles

Performance: Service ceiling 11,400ft at max weight of 53,500lb, 14,800ft at mean weight of 47,000lb; max speed 188mph (206mph if depth charges internal), most economical speed 142mph; climb 11 minutes to 5,000ft at max weight

Armament: One 0.303in in nose turret, 700 rounds per gun
four 0.303in in tail turret, 1,000 rounds per gun
two 0.303in in side positions, 800 rounds per gun (plus 2,000 rounds reserve)
8 × 250lb external, or 4 × 500lb external, or 4 × 475lb DC

(Above) **L2158 was the first production Sunderland I; having been with MAEE on trials, the aircraft went to 204 Squadron in West Africa. The aircraft failed to return from a sortie on 17 August 1942.** Peter Green Collection

One of the first production batch Mark Is, L2163 served with 210 Squadron, as here; 240 Squadron; 10 Squadron RAAF; and 228 Squadron, sinking at Stranraer on 15 January 1942. Peter Green Collection

(Below) **228 Squadron Sunderland being dragged along – muscle power was a frequent feature of the aircraft on land; note the retracted front turret. The squadron re-equipped with Sunderlands in November 1938 and operated various marks through to June 1945.** Reg Parnell

The first two Sunderlands were soon in operation, and along with the remainder of the Squadron were detached on 16 July to the unit's war station at Invergordon, where a period of seven days was spent on exercises with the Home Fleet in the North Sea.

The competing Saunders-Roe design suffered a final set-back in August 1938 when prototype K4773 suffered a wing collapse; development contracts were cancelled and the Air Ministry ordered an additional ten Sunderlands.

Operational Sunderlands were soon involved in semi-clandestine work:

Between 4 and 10 August one aircraft secretly carried Malcolm MacDonald, the Colonial Secretary, from Calshot to Aboukir. The Colonial Secretary was en route to Jerusalem where he made an important announcement on Government policy towards Arab-Jewish problems in Palestine on 6 August. The aircraft returned on 10 August.

Night flying training commenced in September – and soon resulted in the first casualty. On 20 September L2162, flown by Sqn Ldr Watts-Read, crashed on landing, killing two of the crew and injuring four others.

This was the time of the Munich Crisis when it looked very likely that Britain and Germany would go to war; the RAF went on to operational standby and aircraft moved to their war stations. Typical of this was 210's move of its six Sunderlands to Newport on 29 September, the aircraft returning to Pembroke Dock on 8 October with the 'resumption of normal international relations with Germany'. With the Hitler–Chamberlain agreement, encapsulated in the now notorious 'peace in our time' statement, the tension passed and units returned to their peacetime bases. It was generally considered that the war had simply been delayed; the RAF took advantage of the delay to improve its operational capability in terms of equipment, bases and training. HQ Coastal Command confirmed in mid-November that 228

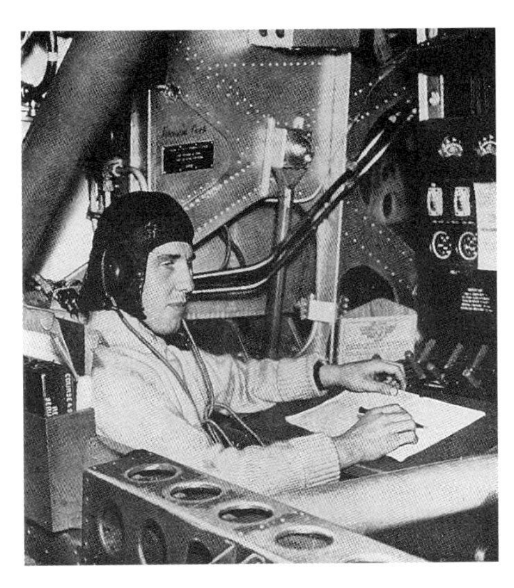

(Above) **A very pensive-looking Flight Engineer.**
Ken Delve Collection

Scanning the sky for enemy aircraft and the sea for enemy vessels or ships in distress, Sunderland crews flew sorties of up to 15 hours' endurance.
Ken Delve Collection

Squadron, under Sqn Ldr L.K. Barnes, would re-equip from Supermarine Stranraers to Sunderlands with an establishment of six IE (Initial Establishment) and two IR (Immediate Reserve) aircraft; the first aircraft, L5805, was collected from Rochester on 21 November. This was not, however, the Squadron's first involvement with Sunderlands: the previous month it, like 210 Squadron, had been involved in the preparation and ferrying of aircraft to the Far East, Flt Lt Laws ferrying L5803.

For the UK squadrons the first months of 1939 brought regular fleet training exercises, typical of these being the late February Trade Protection Exercise 'XKC' in the English Channel and Western Approaches with units of the Home Fleet. One of the primary roles of the Sunderland was to locate and shadow 'enemy' forces threatening 'friendly' convoys. Other exercises involved co-operation with, or operations against, submarines. The early part of 1939 consisted of routine exercises and training, fighter affiliation with various squadrons proving that the Sunderland was quite able to look after itself. Despite its size the Sunderland was quite manoeuvrable, as many a fighter pilot discovered during fighter affiliation exercises; furthermore, it was such a large target that fighter pilots – more used to fighting smaller aircraft – had problems assessing range. With one crewman acting as the Fighting Controller, the Sunderland operated a co-ordinated defence tactic, the Fighting Controller advising the pilot of manoeuvres and identifying targets to the gunners.

The Sunderland force underwent another change in April, 228 Squadron being warned on 26 April that its aircraft would be handed to 202 Squadron, then operating from Gibraltar, whilst it would subsequently re-equip with another type. At Calshot in June 1939, Wg Cdr K. B. Lloyd's 204 Squadron bade farewell to its Saro Londons and on 16 June sent a detachment of pilots and crews to Pembroke Dock to collect the first Sunderlands and also receive instruction from 210 Squadron; meanwhile, the buoys at Calshot were lifted and relaid to give greater distance between them, the extra room being needed by the far bigger Sunderlands. Deterioration of the international situation accelerated in mid-1939 with continued German demands for expansion in Europe. The RAF instigated operational procedures for some units from August, Sunderland units commencing patrols in the Western Approaches from 17 August. On 25 August, 210 Squadron detached two aircraft to Woodhaven and two to the Shetlands (using the depot ship SS *Manela*) to increase the area of cover. With the declaration of war on 3 September the squadrons commenced intensive anti-submarine patrols. Although still primarily equipped with Londons and a small number of Sunderlands, 201 Squadron, as part of No 100 Wing under the command of Wg Cdr Cahill, was also involved with such patrols from bases in Scotland, as well as convoy escort.

'Baggage handlers' 1940 style; WAAF personnel load kit bags into a Sunderland. Ken Delve Collection

This would have made a great caption competition in the 204 Squadron photo album! Ken Delve Collection

CHAPTER TWO

Home Waters and the Atlantic

The Years of Struggle

Outbreak of War

The German invasion of Poland proved the final catalyst that brought France and Britain into conflict with Germany; Britain declared war on 3 September. One of the first operational Sunderland sorties was flown the same day by Flt Lt Ainslee of 210 Squadron: L2165 was airborne at 0500 on a convoy patrol sweeping ahead of seven destroyers that were patrolling the route to Milford Haven. The Squadron detachment in the Shetlands flew four sorties on 4 September. The Coastal Command Order of Battle (ORBAT) at the outbreak of war showed that little had been achieved in terms of expansion and re-equipment, most of the RAF's efforts having gone into its fighter and bomber arms. Of the eleven GR (General Reconnaissance) squadrons, ten were equipped with Avro Ansons; and of the six flying boat units only two had Sunderlands – 204 Squadron at Mount Batten and 210 Squadron at Pembroke Dock.

On 5 September a German U-boat sank the merchant ship *Athenia*, the first such casualty of the war. The U-boat war had started and within 18 months was to assume epic proportions; the Battle of the Atlantic was to be critical to the survival of Britain. No. 204 Squadron flew its first war operation the next day, Flt Lt Harrison taking L5799 on a convoy patrol over St George's Channel in co-operation with six destroyers. The sortie was uneventful until the flying boat returned to base, when it was fired on by a shore battery. No damage resulted in what was to become an all too-frequent example of 'friendly fire'. Despite the fact that the Germans had no aircraft even remotely like the Sunderland, anti-aircraft batteries (and especially naval gunners) seemed inclined to fire at almost any aircraft.

On 8 September Flt Lt Hyde of 204 Squadron was on patrol in N9021 and during this sortie the crew made attacks on two submarines, dropping four bombs on each occasion, but with no observed result. The following day, Flt Lt Harrison bombed a 'submarine feather', an oily patch being subsequently observed near the explosion. A little later, Fg Off Phillips in N9046 attacked a submerged U-boat: 'The aircraft then called on three Polish destroyers to check out the area and these warships made depth charge attacks. Although there was no solid confirmation of a success, C-in-C Plymouth assessed the attacks as having been effective.'

On the same day Flt Lt Ainslee in L2165 of 210 Squadron was airborne from Pembroke Dock on an ASP (Anti-Submarine Patrol) for a convoy and observed the *Empress of Australia* on an evasive zig-zag course. Shortly after leaving the convoy at 1310, the crew 'attacked a U-boat that had just crash-dived and saw a 25-yard patch of oil on the surface after the attack, but no other proof of a kill'. The Squadron recorded four further attacks in September but all without definite result. An aircraft was lost on 17 September, Flt Lt Davies in L2165 having become lost over the Irish Sea and running out of fuel when almost back to safety, the aircraft crashing near Milford Haven with the loss of all on board.

For 228 Squadron the outbreak of war saw five aircraft based at Alexandria and three at Malta, all serviceable and fully bombed up. Having spent a year becoming familiar with the needs of the Mediterranean theatre (*see* Chapter 5), the Squadron had mixed feelings when ordered back to the UK, four more aircraft moving to Malta on 9 September. The following day four Sunderlands flew back to Pembroke Dock via Marignane, although one aircraft had both floats torn off when Sqn Ldr Menzies had to land downwind in poor visibility. During the salvage operation the aircraft turned turtle and was eventually beached in that position. 12 September brought the unit's first war sortie, Flt Lt Smith taking N9025 and Plt Off McKinley L5807 on an anti-submarine patrol to the west of Ireland. Flt Lt Skey made 228's first U-boat sighting on 15 September, but the first attack was carried out on the 24th by Flt Lt Brooks. It is of interest that the Squadron acquired a PBY-4 Catalina 'for operational use' on 18 September; there is, however, little subsequent reference to this aircraft.

There was, understandably, a rash of U-boat sightings and attacks in this early period of the war as crews were keyed up and no-one was sure what was the level of activity of the German submarine service. Inevitably many of these were not true targets: 'A trawler was sighted and almost mistaken for a surfacing submarine', recorded the 204 Squadron ORB; other 'targets' may have included sand bars, whales and submerged rocks.

16 September brought a new Admiralty policy with aircraft being given patrol areas up to 200 miles (320km) out into the Atlantic, listening out on the merchant vessel distress frequency, the idea being to allow them greater freedom during their 10-hour patrol to adopt their own methods of finding and destroying U-boats. The first such operation was flown the following day by 204 Squadron, although the crew had nothing to report; long hours of searching the grey waters of the Atlantic were to be the *modus operandi* of the Sunderlands and other long-range maritime patrol aircraft. On 18 September Flt Lt Harrison (L6799) attacked a diving U-boat and in two attacks dropped two salvoes, each of four bombs dropped in the line of the submarine wake.

> A black object was observed on the surface shortly after the explosion of the second salvo, projecting momentarily from the water close to the position of the explosion. No other evidence of destruction was available, but the C-in-C Plymouth subsequently informed operations room Mount Batten that the submarine was considered sunk. After watching the position for 15

(Above) The Kensington Court **rescue of 18 September 1939, one of the earliest Sunderland escapades to make the news.** Ken Delve Collection

A cartoon view of the Sunderland picking up the ship's crew. 204 Squadron

minutes the aircraft returned to base.' (204 Squadron ORB)

Later analysis meant that this U-boat, like the others mentioned so far, was not credited as destroyed. There continues to be much debate as to the effect of various attacks and the causes for the loss of any particular U-boat; official wartime and post-war records have been frequently queried and it seems unlikely that a 'definitive' list will ever be produced. (The list published here at Appendix IV is compiled from a number of sources.)

Shipping losses to the U-boats meant that rescue operations had to be mounted; one of the first to reach public notice involved the SS *Kensington Court*, torpedoed on 18 September, three Sunderlands of 204 Squadron being involved in a dramatic rescue.

> L5802, Flt Lt Barrett, commenced anti-submarine search at 1215. They made good a track of 240 degrees from Rame Head. At 1337 they intercepted a signal from the *Kensington Court* – 'gunned by submarine' – position 5031N 0827W, 100 miles away from aircraft's DR position. The aircraft made for this position immediately and found *Kensington Court* down by bows, with one Sunderland from Pembroke Dock [228 Squadron] waterborne half-a-mile from the ship. The ship's boat with entire crew (thirty-four) was also observed to be much overloaded, a second boat having been swamped. L5802 carried out a quick submarine search, and then landed, picking up the remaining fourteen men whom the first FB [flying boat] had left. It was learnt that a submarine had opened fire on the *Kensington Court* without giving a warning. The FB saw the ship sink. She carried 8,000 tons of grain.

According to the Coastal Command account, three Sunderlands had arrived at roughly the same time, only just after the crew had taken to their boats. One aircraft circled the area as a U-boat guard while the other two landed to pick up the survivors. The heavy swell meant that crewmen had to be transferred from the boats to the aircraft in rubber dinghies, twenty-one being taken to one Sunderland and thirteen to the other.

Two of the pilots involved in the *Kensington Court* rescue were awarded the DFC – Fg Off Thurston Smith of 228 Squadron and Fg Off John Barrett of 204 Squadron – the joint citation reading:

> Acting Flt Lt Smith and Acting Flt Lt Barrett were, in September 1939, respectively in command of the first and second of three flying boats of Coastal Command which were engaged on patrol duty over the Atlantic when they intercepted messages from a torpedoed merchant ship – the *Kensington Court*. They proceeded to the scene to undertake rescue work. A lifeboat was seen in the vicinity containing thirty-four men and the first aircraft alighted and took on board twenty-one of the men. A thorough search was made by the second aircraft, which afterwards alighted and, in spite of the heavy swell, took on board the remainder of the crew from the lifeboat. (Air Ministry Bulletin 177)

No 204 Squadron recorded a period of intense operations in mid-September with a high percentage of sightings and attacks:

Date	Aircraft	Pilot	Details
Sep 8	N9021	Flt Lt Hyde	attack
	L5802	Flt Lt Barrett	sighting
Sep 12	N9021	Flt Lt Hyde	sighting
Sep 15	N9021	Flt Lt Hyde	attack
Sep 18	L5799	Flt Lt Harrison	attack
Sep 20	N9030	Flt Lt Johnson	attack
Sep 21	N9021	Flt Lt Hyde	attack

Although the Squadron records suggest that at least some of these attacks were successful, none of the attacks was subsequently credited with causing the loss of a U-boat; however, without access to all of the U-boat records it is impossible to assess what effect, if any, these attacks had on their intended victims. Two of the Squadron's aircraft, N9023 (Flt Lt Brooks) and N9020 (Flt Lt Skey), carried out another rescue on 24 September, guiding a rescue ship to survivors from the *Hazelside*, torpedoed by the U-31. Flt Lt Brooks attacked the wake of the U-boat but with no evidence of a result. 16 October saw the loss of a 204 Squadron aircraft, N9030/B crashing on landing at Calshot, four of the crew being killed.

Crews from 10 Squadron RAAF arrived in the UK to collect their Sunderlands (P9600–P9606 and P9620–9624 having been allocated the Australian serials A18-1–A18-9); with the outbreak of war the Squadron was attached to Coastal Command and based at Mount Batten. By September the Squadron had been allocated three aircraft but as it was not operational a number of pilots were loaned to 210 and 228 Squadrons for operational flying. The first true 10 Squadron operation was not flown until 6 February when an escort/ASP (Anti-submarine Patrol) was flown.

The Sunderlands of 228 Squadron were the first to be equipped with dorsal mountings for Vickers K guns, for the port and starboard beams. The gunners had to stand on a platform, their upper bodies being protected by metal windshields. Armament

Pair of 204 Squadron aircraft at Pembroke Dock. Ken Delve Collection

modifications, some official and some not so official, were to be a feature of the Sunderland throughout the war.

Asdic Sunderlands

One fascinating proposal, which in the event came to nothing but which continued to 'bounce' between Coastal Command and the Air Ministry for some time, was for the Asdic submarine-locating system to be fitted to the Sunderland. The initial suggestion was put forward by AOC No. 15 Gp to HQ Coastal Command on 24 September 1939:

> The production of a few flying boats with Asdic gear and depth charges. The boat should carry one or two small (100lb) A/S bombs to keep a sighted U-boat on the move, it should then alight on the water, locate the submarine with Asdic gear and kill it with depth charges.
>
> It is understood that the present M A/S B set Type 130 Asdic gear could be adapted for this purpose. Its present weight is about 1,000lb but it is believed that this weight could be considerably reduced in a set specially made for air work. Even at the present weight a Sunderland could carry the Asdic and two depth charges without reducing its fuel load.

The concept was supported by C-in-C Western Approaches in a letter to the Secretary of the Admiralty, and at a conference held on 30 September it was decided that Captain A/S should consult Shorts about design and fitting of such equipment. Coastal Command was not, however, convinced and a memo of 4 October put powerful counter-arguments covering such items as the need for a calm sea (only twenty days a year), vulnerability to gunfire from the U-boat, and concluding that 'it would be most interesting to carry out trials with a flying boat towed behind a High Speed Launch at 15–20kt, and see the result of releasing depth charges on the structure of the flying boat'. An Air Ministry conference of 10 October concluded that 'the scheme was neither practical nor worthwhile'.

AOC No. 15 Gp refused to let go the idea and the following August re-submitted the proposal; however, the official records make no further mention until 3 July 1941, a Coastal Command conference on Asdic concluding:

> The matter is not one of extreme urgency but he [C-in-C Coastal Command, ACM Sir Philip Joubert] would like Shorts to look into the possibilities and advise on:
>
> 1. what is the engineering problem.
> 2. how serious would be the modification.
> 3. if effectiveness of equipment could compensate for loss of labour involved in fitting it to Sunderlands.

This conference earned the C-in-C a rap on the knuckles from the Chief of the Air Staff as no one from MAP had been at the meeting. Again the records go silent, and it seems that the Asdic Sunderland never made it beyond the drawing board.

Coastal Command's air resources were over-stretched and this was particularly true of its maritime patrol aircraft. This meant that detachments had to be sent to various bases as and when required, this often involving no more than a single aircraft. Typical of this was N9023 of 228 Squadron deploying to Oban on 24 November for 'operational duties' in connection with the search for the German pocket battleship *Deutschland*. The aircraft also operated from Sullom Voe and Invergordon, and during its deployment conducted a number of convoy patrols in addition to shipping searches. Operating under No. 18 Group, 210 and 228 Squadrons had been deploying aircraft to Oban, Invergordon and Sullom Voe in the later months of 1939, a commitment that continued into 1940.

There was pressure to increase both the operational load and range of the Sunderland to meet the ever-increasing operational task. Sqn Ldr Pearce of 210 Squadron took an 'Air Ministry official' on All-Up Weight trials with L5798 on 2 November:

> Take-off tests at 56,000lb; attempts made to get on step downwind and then make a turn into wind while still on step – not a success and the airscrews of the inner engines bent making the aircraft unserviceable.

The notional endurance of the Sunderland I was in the order of ten hours, but crews were soon exceeding these times. In common with other large aircraft, the Sunderlands were instructed to carry homing pigeons:

> Pilots were given a lecture on the management and handling of pigeons, a scheme having been initiated whereby all aircraft would carry homing pigeons on operational flights, for emergencies such as wireless breakdown. (204 Squadron ORB, December 1939)

To increase Sunderland production a second line was set up at Blackburn aircraft, in conjunction with the Denny Bros shipyard at Dumbarton, on the Clyde.

As the winter weather of 1940 arrived, two new policies were brought into force for the UK squadrons: firstly, the standby flying boat was to be started up and taxied in order to keep it at readiness; secondly, each aircraft was to be brought up the slip once a week, or after six hours spent in ice-forming conditions, to be covered with Kilfrost de-icing paste. Coastal Command instructed that at least two boats per squadron were to be kept prepared to fly in conditions 'likely to produce ice formations'. On 1 January, No. 204 Squadron at Mount Batten recorded that it had five aircraft serviceable and 'ready to come up the slip for float modifications by the makers and the application of Kilfrost'. There were a number of other references to Kilfrost the same month. On 6 January, Sunderland N9021 was brought up the slip for the application of camouflage and Kilfrost, work which was completed on the 14th. On 18 January the Squadron recorded that

> N9046 [one of the aircraft given Kilfrost on the 1st] had to return to base owing to being unable to get through a front owing to snow and ice accretion – the Kilfrost was outworn (sic) since it had been applied more than a week before. The aircraft would have had to return apart from this for fear of ice accretion on instruments and carburettors. (204 Squadron ORB)

At the same time, the Squadron was conducting trials on an autopilot, the first three aircraft (N9021, N9044 and N9047) having been equipped by 5 January. 'None have yet been tested as all are awaiting certain modified parts which were on demand, to be exchanged for parts on the autopilots installed.' The first autopilot test was flown on 9 January by Flt Lt Falkener from the Air Ministry and Mr Fen from Farnborough, who declared the installation 'serviceable for operational purposes after a few minor adjustments'. However, by 21 January the ORB was recording that:

> After the various tests of the three Farnborough autopilots carried out this month, the general opinion of these amongst pilots of 204 Squadron was that they were verging on being useless owing to excessive precession. The rate of precession was reported by Flt Lt Hyde to be 30 degrees in five minutes. At the same time, the experts who had flown on these tests considered the operation satisfactory.

Three days later Fg Off Johnson flew N9044 on an evaluation with two squadron leaders from the Air Ministry; it was recorded that the latter were not impressed as the rate of precession was approximately 84 degrees per hour. From that point on, the 204 Squadron ORB is silent on the matter of autopilots.

For 228 Squadron the early part of the year saw operations from their main base at Pembroke Dock but also detachments to

From mid-January 1940 concern was being expressed as to the future development of flying boats for the RAF. On 13 January, C-in-C Coastal Command wrote to the Air Ministry (AM Sir Wilfred Freeman):

> I am very concerned about the abandonment of the flying boat development programme and the decision to cancel Specification R5/39. I understand that for the moment, all that is being done is to improve the performance of the Sunderland and to increase production of this type. You'll be the first to admit, from our long and bitter experience in increasing the performance of aircraft by installing more powerful engines, it has never proved successful.
>
> Whilst, possibly, the modification proposed for the Sunderland may give this aircraft a further 400 miles range when loaded up to the maximum permissible take-off weight, at most this will only result in an operational range of 1,800 sea miles with full bomb load or 2,000 sea miles with no bombs, whereas I can visualize in three months' time, our flying boats being called upon to operate for longer than twenty hours at a time or for instances greater than 3,000 sea miles. So, unless we are going to purchase aircraft from America, we shall be unable to fulfil this requirement.
>
> While I agree that it would be a waste of time to proceed with the R5/39 Specification, I do definitely feel that we should continue with our flying boat development.

Bowhill continued his letter with an outline of a specification for a flying boat to be powered by four 1,600hp engines, an all-up weight of up to 84,000lb, with adequate bombs and defensive armament. He suggested that Saunders-Roe could adapt their R5/39 drawings.

Two days later he wrote to Freeman again:

> I understand that you are contemplating purchasing a certain number of PBY4s from America, and that for reasons of conserving our foreign exchange you have decided against the four-engined PBY Model 29 in favour of the two-engined PBY model 28 modified to our requirements. My objection is that at best this aircraft does not compare as a military proposition with Sunderland or Lerwick in armament or navigation facilities, but where the Americans do definitely have an advantage over us is in their engine design.
>
> I have been investigating the possibility of installing American engines of similar power to the Pegasus XXII in these [Sunderland] flying boats. I consider that if the Pratt & Whitney R-1830-S3030 were installed in Sunderlands and extra tankage fitted to bring fuel capacity to 2,500 gallons, the performance would be increased to that of the PBY-4. By this method we should obtain a military flying boat designed for our needs, and at the same time save a considerable amount of foreign exchange.

A memo circulated on 19 March included a copy of a paper entitled 'An aspect of flying boat plans' first issued on 19 December 1939; this contained a number of interesting statements such as 'The Air Staff have considered that the Sunderland will meet all requirements in the air at sea for the duration of this war.' But, as we have seen above, within months of this, C-in-C Coastal Command was by no means convinced of this. The paper went on:

Sunderland I L5802 of 204 Squadron at Rochester. Andy Thomas Collection

> Consideration of the immediate requirements, i.e. the modernization of the Sunderland, logically leads up to its replacement. Air Staff has agreed to increase the range rather than increase the speed of the Sunderland. The added fuel capacity will increase the maximum all-up weight from 51,000lb to 56,000lb. At the latter all-up weight it has been found that the inboard airscrews are liable to get damaged when taking-off. Tests now show that a slight modification to the forebody of the hull would overcome this.
>
> The following are, therefore, the modifications proposed for the Mark II Sunderland:
>
> a. Fuel tankage increased by 500 gallons.
> b. Forebody of hull modified.
> c. Bomb gear redesigned so that the gear can be power operated. This will reduce the time of getting the bombs from the stowed position inside the centre section of the hull to the operating position under the wings from ten minutes to ten seconds. On account of the time taken to get the bombs into position, the Sunderlands are being operated with the bombs carried externally, which reduces the cruising speed by 15mph with consequent reduction in range.
> d. Folding wing tip floats should increase the cruising speed by about 4mph and add a little to the range.
> e. Twin fins in place of existing single fin.
>
> Under normal conditions a replacement aircraft should be started immediately a new aircraft enters service. A Sunderland replacement should have been put in hand over twenty months ago.

A meeting was convened by DGRD (Director General of Research and Development) on 24 April to discuss a Sunderland replacement, a six-engined flying boat and a small flying boat. Only the Sunderland discussion concerns us:

> It was pointed out that the Sunderland came into service early in 1937 when a new Specification should have been put in hand immediately. Had this been done a Sunderland replacement would have been coming into service this year or in 1941. The decision to order a replacement for the Sunderland was not given until July 1939, but when war was declared the order was stopped. It was necessary that an order for a flying boat to R5/39 Specification should be placed with Shorts and Saunders using Centaurus or Sabre; this would give a military aircraft that was also readily adaptable for civil use.

Discussions continued in various fora, but progress was slow and there was by no means a consensus of opinion. On 16 August, Bowhill wrote to AVM Douglas (DCAS):

> I have been told that they are discussing the armament required on the new flying boats, Specification R13/40 and R14/40, and I suggest that either I should be consulted or that I should be allowed to send one of my officers on this important matter. After all, we are the people who have to fly and fight these boats, and again I consider Coastal Command knows just as much, if not more, about flying boats as most people, yet apparently we are always studiedly ignored.

Douglas replied two days later that he had passed the comments to Saundby as ACAS(T) as it was his area of responsibility. In a memo of 21August Bowhill was invited to send a representative to the next meeting. Little actually happened until early 1941, before which Bowhill had asked Sir Henry Tizard for an update on the projects; Tizard replied on 17 January 1941:

> The R13/40 is still going on, but it is under a bit of a cloud. I myself am very doubtful whether it is in the least likely to be of any use to you in the present war. As for R14/40 the estimate is that even if everything went well you would not get any appreciable number for operational use within two years. The developments have not been stopped but the Minister [of Aircraft Production] is varying existing contracts. I am doing my best to get the two firms concerned to co-operate instead of compete.

A meeting was held at the Ministry of Aircraft Production on 18 February to discuss flying boat policy. It was a full and frank exchange of views and amongst the pertinent comments Bowhill stated that

> ... the best, if not the only, counter to the U-boat menace is the flying boat. As regards the suggestion that the landplane would serve the purpose equally well, the reasons for the flying boat's superiority are too many to enumerate. Apart from better navigation facilities, and greater comfort offered by flying boats, there is the important psychological factors to bear in mind. There is an urgent need of an aircraft capable of flying 1,000 miles out to sea and operating at this distance from land for hours on end. The strain on crews asked to make such flights in landplanes would be enormous and could not be accepted.

It was agreed to proceed with the R14/40. It was also agreed that the Saunders-Roe project be dropped and that Shorts be asked to provide two prototypes to R14/40 and be given an initial order for 25–30 aircraft, the firm to jig and tool up for production of two aircraft per month.

The effective result of the delays and cancellations was that the Sunderland had to soldier on and, if possible, be further developed.

Convoy protection was a major part of the work carried out by the maritime squadrons. Ken Delve Collection

A camouflaged 10 Squadron RAAF Sunderland lies close inshore at Mount Batten along with a white-painted aircraft from another unit. Ken Delve Collection

Invergordon, the latter usually by a single aircraft tasked to operate with a particular convoy. This routine of deploying aircraft to bases in Scotland continued for much of the early part of 1940. The first credited Sunderland success against the U-boats, albeit shared with the Royal Navy, went to Flt Lt Brooks (N9025) of 228 Squadron, U-55 being sunk on 30 January. Having found the submarine on the surface, Brooks bombed and strafed it before directing surface ships to the position. The German crew subsequently scuttled the boat and survivors were picked up by a RN vessel. Surprisingly, the Squadron records make little mention of this attack and what was, for the Sunderland, a historic first victory has been given little detail in the historic record.

With effect from 1 March aircraft were ordered to take a fuel load of 2,000gal (9,100ltr) rather than the previous 1,700gal (7,700ltr) as longer daylight hours meant longer patrols. It was a reasonably busy period for the Sunderland units, although this seldom meant more than one operational sortie a day, 204 Squadron being typical (*see* table right).

30 January 1940; this U-boat was attacked by Flt Lt Brooks of 228 Squadron and was subsequently scuttled by its crew. Ken Delve Collection

Sullom Voe in April 1940 and Sunderland I L5805 of 201 Squadron awaits its next mission; the aircraft subsequently served with 95 Squadron and was lost in October 1942. Andy Thomas Collection

Date	*Aircraft*	*Pilot*	*Time*	*Details*
Mar 1	N9044	Flt Lt Hughes	0754–1920	Convoy
Mar 2	N9028	Flt Lt Gibbs	1050–1840	Convoy (not found)
Mar 3	N9047	Flt Lt Harrison	0740–1750	Convoy (not found)
Mar 4	N9046	Flt Lt Phillips	0725–2045	Convoy
Mar 5	N9024	Flt Lt Hughes	0725–1510	Convoy
	N9028	Flt Lt Gibbs	1020–2040	ASP*
Mar 7	N9046	Flt Lt Phillips	0740–1940	Convoy
Mar 8	N9028	Flt Lt Gibbs	0750–1955	Convoy
Mar 10	N9046	Flt Lt Phillips	0720–2035	Convoy
Mar 11	N9021	Flt Lt Gibbs	0630–0850	
Mar 12	N9028	Flt Lt Gibbs	0725–0920	Convoy**
Mar 15	N9047	Flt Lt Harrison	0721–2016	Convoy
	N9028	Flt Lt Gibbs	0658–2000	Convoy
Mar 19	N9028	Flt Lt Gibbs	0702–1957	Convoy
Mar 20	N9047	Flt Lt Harrison	0707–1959	Convoy
Mar 21	N9046	Flt Lt Phillips	0503–2021	Convoy
Mar 22	L5802	Fg Off Johnson	1155–2010	Escort aircraft carrier
Mar 23	N9024	Flt Lt Hughes	0645–1945	Secret convoy
Mar 25	N9046	Flt Lt Phillips	0625–1722	Convoy
Mar 27	N9028	Flt Lt Gibbs	1542–2000	Convoy
Mar 28	N9028	Flt Lt Gibbs	0625–2015	Convoy
Mar 29	N9024	Flt Lt Harrison	0355–1120	Convoy (not found)
	N9044	Fg Off Harrison	1408–2230	Convoy
Mar 31	N9044	Flt Lt Phillips	0642–1320	Convoy

** A report from Destroyer 62 said that it had attacked a submarine but the aircraft could find no trace; at 1820 the Sunderland bombed a large patch of oil but only one of the eight bombs exploded.*

***Returned to base early due to poor weather.*

There were still many changes being made to the aircraft and 204 Squadron was complaining that it was still waiting for gunsights and cameras, and that the 'self-operating' turret was causing problems as it had a tendency to 'creep'. Indeed, weapon trials were very much part of this period, with depth charges and 500lb anti-submarine bombs being evaluated; as early as the end of 1939 there had been complaints that the aircraft's standard offensive weapon, the 250lb A/S bomb, was unreliable and ineffective. On 8 March Sqn Ldr Thomas in L5802 of 304 Squadron carried out a trial drop with four 500lb AS bombs fitted with 7-second delay fuzes, in the vicinity of the Eddystone lighthouse, the results being assessed as satisfactory.

Squadron records often contain obtuse references to equipment or operations; the 210 Squadron entry for Mar 14 states that 'Wg Cdr Fressanges, Sqn Ldr Pearce and a passenger tested secret equipment fitted to N9027.' This is almost certainly a reference to ASV (Air to Surface Vessel), the radar equipment that was to prove a vital part of the maritime aircraft's war against the U-boats.

Invasion of Norway

Norway had become the focus of attention in April, crews being tasked to monitor German naval activity but also to reconnoitre the coast to gain intelligence for possible future operations. 1 April saw 204 Squadron complete its move to Sullom Voe, four aircraft having been flown to Pembroke Dock for hand-over to 201 Squadron, thus leaving Mount Batten free for 10 Squadron RAAF, who moved in during April. April also saw 201 Squadron say farewell to the last of its Saro Londons – and eagerly await more Sunderlands, although the flying boats were in short supply despite the additional production facility at Dumbarton.

The 250lb anti-submarine bomb was the standard weapon for the Sunderland in the early part of the war; this weapon proved remarkably ineffective. Ken Delve Collection

Anti-Submarine Bomb

Anti-submarine bombs were developed as a result of the British experience against German submarines during World War One. In the 1920s two basic A/S weapons were under development – 250lb and 500lb bombs – although the Admiralty requested a 100lb version. All three were subsequently developed but the 250lb A/S bomb was the one ordered, in the mid-1930s, in large quantities. By the outbreak of war the bombs were already being assessed as being of

> ... doubtful quality and untried on operations, with unreliable fuzes. Moreover, the underwater path of the bombs was unpredictable, particularly when released from low altitude and because of the small lethal range of the charge extremely accurate bombing was essential. Such accuracy was difficult to achieve as no reliable low level sight then existed. (SD719, Armament Volume I)

The A/S bomb certainly attracted numerous adverse comments during the early part of the war and the development of the depth charge was seen as one solution.

On 31 March 1942, the C-in-C Coastal Command wrote to the Air Ministry:

> For the entire duration of the war to date, we have been attempting to attack U-boats with two types of depth charges, neither of which is capable of giving satisfactory results. We started off with the standard Admiralty depth charge and have made attempts to make it a satisfactory aircraft weapon. Despite all attempts to improve it, it has been found impossible to clear the Mk VII DC for heights above 150ft and speed above 150kt. [See below for depth charge development]

With reference to the 250lb Mk VIII the letter stated:

> This weapon has been more satisfactory from the point of view of tactical use since it is not restricted in height and speed, but its killing power is quite inadequate and is seriously impeding the effect of our anti-submarine war as a whole. We therefore request that a special depth charge be developed. (SD719 Armament Volume I)

This requirement was in effect turned into the 600lb A/S bomb and as this weapon, when developed, was not allocated to the Sunderland, then it does not concern our story.

Depth Charge

The most effective weapon in the anti-U-boat armoury was the Depth Charge (DC). In the light of experience, and as the result of continuous trials and development, changes were made at various times to the standard stick spacing to be used with DCs. The recommended spacing changed from 60ft (18m) to 36ft (11m) and, finally, at ORU instigation in 1943, to 100ft (30m). As the primary weapon used by the Sunderlands in their fight against the U-boats, the depth charge is worthy of detailed examination.

By early 1940 there was mounting evidence that the existing anti-submarine bombs were ineffective and Coastal Command began looking for alternatives – the most obvious of which was the depth charge, a standard naval weapon for use against submarines. This weapon would have a number of advantages: it had a simple hydrostatic pistol, a high charge-to-weight ratio, and it could be dropped from a low height without the danger of self-damage from the explosion (although in the event this was not always to prove the case). Air-drop trials had been carried out in late 1939 by the Torpedo Development Unit (TDU) at Gosport with an adapted naval 450lb depth charge, and although a number of problems were revealed, the weapon showed promise. In March 1940 the TDU was developing a more suitable version and this 'rough and ready bomb' was tested from a Sunderland off the Isle of Wight on 16 April 1940:

> [It was] found to fall and to detonate satisfactorily and was immediately adopted as a successor to the anti-submarine

bomb. It was not within the power of the Command to convert unlimited numbers of depth charges, and in June 1940 the work was undertaken by MAP to whom drawings of the conversion parts, nose, tail and suspension band were sent. (SD719 Armament Volume I)

No. 204 Squadron had been involved in these trials since December 1939 but the 16 April drop was made by a 10 Squadron RAAF aircraft, this unit having taken over the task. The Air Historical Branch Narrative on the 'U-boat War' is at variance with SD719 in that it states that the April trial was with a dummy weapon to decide the best shape for the depth charge and that more work was needed to develop nose and tail units to get the best ballistic results. A conference was held by No. 17 Group at Gosport on 24 April; the Air Council was not keen to proceed with the project but C-in-C Coastal Command persuaded them to agree to more trials. On 4 June a 10 Squadron aircraft carried out the first full-scale live trials. Later in the month the Command placed an order for eight Mk VII depth charges for operational evaluation by the Squadron. The first operational drop was made on 7 July, the target being a 'suspicious patch of oil': both depth charges functioned perfectly. A few weeks later, on 31 July, the first attack was carried out on a definite U-boat and on 16 August the first credited result was achieved, a 210 Squadron Sunderland damaging U-51.

By August Coastal Command had acquired for conversion a further 700 depth charges, enough to cover three months of operations. These were still in many respects 'interim' weapons, the first significant improvement being the incorporation of a pistol that would allow the depth charge to be dropped 'safe' in the event of a problem – the previous pistol would explode when the depth charge reached the depth setting and if the Sunderland had ditched with the weapons still aboard they would go off with the very real risk of killing any survivors from the ditching (there are various references in this book to the loss of aircraft and crew from such instances). The Mark X pistol had a pin which could be inserted for the appropriate depth setting, the pin acting as a stop for the arming spring; as the pin was not removed until the depth charge actually left the aircraft, the normal condition of the weapon was 'safe'. The only other significant improvements were to the suspension and carriage system, although various changes were made to the pistol in subsequent months.

As the 450lb depth charge was unsuitable for aircraft such as the Swordfish, Walrus and Hudson, a 250lb version was developed, entering service as the Mark VIII. However, by mid-1942 an improved version, the Mark XI, had become the standard Coastal Command depth charge, this having a concave nose in order to impart instability on water impact and make the depth charge travel broadside on, thus making it more effective against its intended target.

This latter point was important in order to give the weapon increased effectiveness against U-boats; however, even more important was range of depth settings, it having been decided that the original 50ft (15m) minimum setting was too deep. After numerous trials, the Mark 16 pistol was introduced with a minimum setting of 22ft (7m); this pistol, teamed with the Mk XI depth charge, became the standard air-dropped weapon for the remainder of the war. The other significant change in 1942 was the use of Torpex as the explosive content, this being some 30 per cent more efficient than the Amatol explosive previously used.

When the U-boats changed tactics and remained surfaced in order to fight-off air attack, aircraft losses increased; in mid-1943 the anti-U-Boat Committee decided that the maximum height for release should be increased to 1,500ft (500m) and speed to 250kt (460km/h), although it was late 1944 before the Mark XIV, designed to meet these parameters, entered service.

Sunderland P9600 of 10 Squadron RAAF leads a rescue ship to a dinghy full of survivors, 23 June 1940. Ken Delve Collection

The installation of Air to Surface Vessel (ASV) was to transform the capabilities of aircraft in the war against both ships and submarines: 'ASV equipment in Sunderlands has made it possible to do more work with convoys at night than has hitherto been regarded as practicable or safe.'

The first prototype Sunderland Long Range Air to Surface Vessel (LRASV) installation was carried out in September 1940; the system used beam arrays similar to those of the Whitley aircraft used for the original trials, with an additional forward-looking dipole array, but using the ASV Mk II frequency of 176 megacycles. The Sunderland prototype was ready in December but initial trials were disappointing, with the best result being a 5-mile (8km) maximum range – and that against a coastline rather than a U-boat. According to the RAF post-war volume 'Signals – Radio in the maritime war',

ASV III was the first effective radar system fitted to the Sunderland; the aerials along the top of the fuselage were very distinctive. Peter Green Collection

> Adequate prior consideration had not been given to the new problems of installation in the flying boat, with the result that improvised mechanical and radio fitting was inefficient. Investigation revealed that certain radio and engineering faults would have to be eradicated before any improvement in performance could be obtained. It was therefore decided to improve the aerial matching, to replace the Sterba arrays by stacked horizontal dipoles, and to install a Yagi-type homing aerial in place of the forward-looking dipole system.

It was also intended to provide an indicator, based on a 4in Cathode Ray Tube (CRT), to be installed in the blind-flying panel at the second pilot's position; however, there were problems with developing a suitable indicator and it was not until early 1941 that Telecommunications Research Establishment (TRE) designed and developed a suitable indicator.

The Coastal Command Development Unit (CCDU) was formed in October 1940 at Carew Cheriton to conduct Service trials, and to examine and develop tactical employment of existing and new radio devices – such as ASV. The Unit had one of each Coastal Command aircraft types, the flying boats being based at Pembroke Dock. (The CCDU was eventually absorbed into the Air/Sea Warfare Development Unit in March 1945.)

Trials were carried out by the CCDU in the early part of 1941.

It was recommended that a qualified ASV operator should be included in the crew; this led to the creation of a new aircrew designation – WOP/AG (Coastal). In February 1942, AOC Coastal Command recommended three such qualified individuals be included in each crew.

Other detection systems were also under development. In February 1941 the CCDU conducted trials on an experimental prototype of a Magnetic Detection System (MDS), developed by Professor E.J. Williams. However, it had a maximum detection range of only 200ft (60m) and as this was the same as Coastal Command's minimum 'operational requirement range' the system appeared impracticable. Nevertheless, such was the need at this time for any detection system that would help in the battle against the U-boats, installation of the MDS Mk II was underway on production-line aircraft by April. Within weeks the decision was changed again and because of the 'limited tactical value' of the system, development and installation was suspended. In spring 1942 MDS was once again under development, although the Sunderland was no longer part of the programme, the Whitley being chosen as the operational platform.

CCDU received the first fully-equipped LRASV Sunderland from Shorts in May 1941, and following adaptation of the feeder system, satisfactory results were obtained, the direction-finding properties of the forward-looking arrays proving particularly accurate and effective. The following month Shorts, and other contractors, began delivery of ASV-equipped aircraft.

Meanwhile, a modification programme was started to fit ASV to some thirty Sunderlands already in service. However, although the system was notionally operational it often required up to 100 hours of work by radio fitters to bring LRASV installations up to a fully serviceable condition. By December there was disquiet at Coastal Command HQ regarding the installation layout – although HQ staff had agreed to the compromise layout that was fitted to the aircraft. There was also concern over the effect of the aerial arrays on aircraft performance.

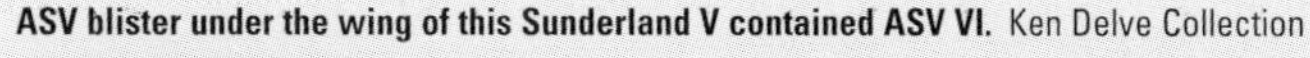

ASV blister under the wing of this Sunderland V contained ASV VI. Ken Delve Collection

No. 204 Squadron flew its first mission from its new base on 3 April, Flt Lt Phillips in N9046 getting airborne at 1125 for a convoy escort. It was to be a hectic start as the aircraft was engaged by six Ju 88s off the coast of Norway. The Squadron ORB recorded the incident:

> Aircraft located the convoy in the estimated position. At 1550 while the aircraft was twenty miles away from the convoy on the starboard bow, two enemy aircraft were sighted low on the water approaching our aircraft from the direction of the Norwegian coast. Both our aircraft and the enemy were 50–100ft above the sea. The enemy circled around our aircraft for two minutes, and then carried out a beam attack on the starboard side at a range of 800yd, the two aircraft flying past successively on a course parallel to our own. A desultory engagement took place without any apparent damage to either aircraft. Both turrets and the starboard amidship gunner fired short bursts. The subsequent appearance of four more Ju 88 aircraft led to the belief that the enemy was employing this tactic to draw our fire.
>
> After three minutes' fighting the two aircraft began to climb turning away from our aircraft, and after a three-minute pause, four more aircraft appeared from the same direction, and immediately delivered a line astern attack on our tail. The rear gunner, Cpl Little, held his fire till the leading aircraft was at 100yd range. He then opened fire and shot down the leading attacker, which banked steeply and dived into the sea. The second aircraft was also hit and was afterwards known to have force-landed. During this attack the first two aircraft to arrive were attempting to bomb the Sunderland from a height of 1,500ft. The bombs however were easily avoided by the fire control officer in the observation dome, directing the pilot away from them. The bombs could be seen falling as soon as they were released, and the nearest one burst at least 100yd from our aircraft.
>
> As soon as the attack of the four aircraft had been broken up by our gunfire, the five remaining aircraft immediately made off, and our aircraft made for home, having suffered the following damage: bullets in port inner, starboard middle and starboard inner fuel tanks causing loss of 500gal during return flight; bullet holes in hull, fin, instrument panel; trimming gear and fuel jettison systems made unserviceable; one bomb rack damaged; the fire control officer and navigator both sustained face cuts from splinters. (204 Squadron ORB, April 1940)

The Sunderland defended itself well, claiming one enemy fighter shot down and one damaged; the flying boat was hit in two fuel tanks and the control wires and certain instruments were shot through. The action brought a DFC for Frank Phillips and a DFM for Cpl William Little, the citation reading:

> The above awards are for gallantry in action. Flt Lt Phillips and Cpl Little were captain and rear gunner of an RAF aircraft attacked by six Ju 88s when on convoy patrol. By his skilful handling of the aircraft Flt Lt Phillips avoided the first attack by machine-gun fire from two enemy aircraft and manoeuvred so that Cpl Little, by well-directed fire which in the face of oncoming attacked he withheld until the enemy were within 100yd, was able to bring down one of the enemy aircraft into the sea in flames. This determined attack terminated the action, the remaining enemy aircraft retiring at speed. (Air Ministry Bulletin 534)

On 6 April Flt Lt Phillips flew the damaged Sunderland from Sullom Voe to Pembroke Dock for repair, having to land at Invergordon to refuel as the aircraft could only carry 1,000gal of fuel because of the damaged tanks.

The coast off Norway remained hazardous and on the 7th 204 Squadron's Flt Lt Harrison (L5799) failed to return from a reconnaissance patrol in this area. The following day the CO, Wg Cdr Hyde, was on a reconnaissance of the Trondheim Fjord when he was attacked by two enemy aircraft, damage being inflicted on the Sunderland before it was able to escape. Flt Lt Hyde crawled into the wings and stopped up holes in the fuel tanks, thus enabling the aircraft to make it back to base. The long-anticipated invasion of Norway was now less than twenty-four hours away.

At 1400 hours on 8 April a Sunderland sighted a convoy of German warships 130 miles (210km) from the Alsboen Light off the west coast of Norway:

> The ships saw the Sunderland almost at the same moment and opened anti-aircraft fire which was both heavy and accurate. The Sunderland was hit almost at once; two of its tanks were holed and the hull gradually filled with petrol. When it landed at its base it had lost 300gal. (*Coastal Command*)

The aircraft had spotted part of the German invasion force on its way to Norway. Throughout the day various Coastal Command aircraft were tasked with monitoring German activity – 'to observe, shadow and report movement of enemy sea forces' – in the area, a routine that continued over the next few days. The following day British forces landed at Namsos as part of a plan to cut Norway in half by taking control of Trondheim, whilst other forces landed to the north at Narvik. Both expeditions were to prove ill-fated.

It was not long before German capital ships were using the sheltered Norwegian fjords, and part of Coastal Command's routine task was the monitoring of likely hiding places. The Admiralty maintained a plot of all the major warships and grew concerned when any 'went missing'. 228 Squadron recorded a typical reconnaissance sortie on 10 April:

> L5806 Flt Lt Brooks having detached to No. 18 Group in the Shetlands flew an afternoon reconnaissance of Trondheim Fjord to report enemy forces there. An aircraft of 204 Squadron had been the previous day and been driven off by enemy aircraft before completing the mission. It was therefore decided to make use of a cloud cover approach from the north, inland and make the reconnaissance whilst flying in a westerly direction. Landfall Vikten Island 1600 and aircraft proceeded inland above broken cloud at 7–9,000ft. Through cloud gaps vessels were seen at anchor in harbour, identified as *Hipper* class cruiser, two destroyers and three large MVs. To NE of harbour was lying the *Nürnberg* and in a fjord N of the town, one destroyer. Engines were de-synchronized and the aircraft constantly retired into cloud. Considerable icing was experienced and heavy snow storms were encountered. Set course due West at 1750 and land at Shetlands 1955. (228 Squadron ORB, April 1940)

Two days later the *Nürnberg* was located again, this time accompanied by the pocket battleship *Lützow*, both of which opened fire at the Sunderland. Flt Lt Van der Kiste of 210 Squadron was airborne from Sullom Voe on 12 April at 0500, tasked to look for troopships in the Hardangar Fjord:

> Heavy snowstorms over the Norwegian coast … whilst searching for the objective, heavy anti-aircraft fire was encountered over Hagesund. The starboard middle tank was hit and the hull pierced. The search continued and the aircraft was hit again over Garvin, the port middle tank and tailplane being hit. The ships were not located. When the aircraft landed at Sullom Voe at 1130 it had 400gal of petrol in the bilges. (210 Squadron ORB, April 1940)

From 13 April onwards each Sunderland carried a Norwegian crew member on these reconnaissance flights to take advantage of their local knowledge. One of 201's aircraft had the distinction of aiding an escape:

> Sighted a fishing boat off the coast of Norway – at first they thought it was laying mines. As the Sunderland approached, a circling Dornier fled – at which point the fishing boat hoisted a Norwegian flag and waved to the flying boat: it transpired that the boat, which then headed to a safe landing in England, carried the British Vice-Counsel in Stavanger, his family, and an ADC to King Haakon of Norway. (201 Squadron ORB)

15 April saw 228 Squadron's CO, Wg Cdr Nicholetts, tasked to fly General Carton de Wiart to a rendezvous with a Tribal-class destroyer in Namsos Fjord. The Sunderland landed in the fjord at 1645 – just in time to see the destroyer under attack from four Ju 88s and two He 111s, a number of which then machine-gunned the flying boat, 2nd Lt Elliott being wounded. One enemy aircraft dropped a long stick of bombs ahead of the manoeuvring Sunderland but the pilots managed to evade this attack, N6133 suffering no significant damage. Sunderlands of 228 Squadron continued their involvement with transporting 'key personnel' to Norway – sorties which often proved hazardous. Sunderland N9025 landed in the fjord near Molde on 27 May and was promptly bombed by twelve Ju 88s, although the only damage was splinters in the tail; however,

> At 1725 the aircraft had to take-off as the engines had become very hot after continual avoiding action. Once airborne the aircraft was attacked by Me 110s from astern, no damage was caused and the gunners claimed one enemy aircraft. (228 Squadron ORB, May 1940)

The Invasion of France

The German invasion of Belgium and France on 10 May brought a heightened state of alert: 'Owing to the state of

Operating the Sunderland

Ken Robinson's book *Dice on Regardless* (R. J. Leach & Co. London 1993) provides a very good insight into Sunderland operations in West Africa and into the aircraft itself:

> When the crew were ready we were all driven out to the aircraft in a motorized dinghy. As we approached the flying boat, I was once again astounded at the size of it. The huge plump hull loomed out of the water to a height of between 12–15ft, excluding the fin. The wing area seemed enormous.
>
> Once the skipper and I had boarded, we stood aside in the bow compartment to allow the rest of the crew to get aboard and go about their various duties. We had to bend down to avoid banging our heads on the top of the nose section, but I was soon to discover that this was the only compartment except the rear turret where this precaution was necessary for anyone of medium height. Luxury indeed.
>
> The front turret had been withdrawn into the bow compartment to enable the crewmen to moor and unmoor the boat. There was a bollard, hinged at the bottom, which was normally horizontal when the aircraft was in flight but was now fixed in an upright position adjacent to the bomb-aimer's hatch. Strapped to the hull on the port side were a mooring ladder and an anchor for use when there were no suitable fixed mooring facilities. On the starboard side there was a J-type inflatable dinghy and alongside a companionway leading to the flight deck was the toilet compartment, complete with lavatory and wash-basin. Just aft of the forward hatch through which we had boarded was the entrance to the wardroom and a through route to the tail.
>
> Moving towards the rear of the aircraft, we passed through the wardroom which contained two bunks doubling as seats on either side of a table with hinged flaps. Further aft was the galley. The equipment included two primus stoves separated by a small oven. Another companionway ascended to the rear end of the flight deck and on either side of the galley was a square hatch used, among other things, for deployment of the drogues, which were stowed in semi-circular receptacles just below the hatches.
>
> Beyond the galley was the bomb bay in which eight 250lb depth charges were normally suspended from racks. These racks would be run out along the underside of the wings when the DCs were made ready for an attack. Under the port bomb door was a removable wedge-shaped panel which enabled a spare engine or other sizeable cargo to be carried. The next compartment had two more bunks in it and then there was a very spacious area leading up to the tail of the aircraft and the rear turret. The equipment in this space included a fitter's bench and there was a hatch on the starboard side providing an exit at a higher level above the water line than the forward hatch.

Wing Commander T. Holland flew Sunderlands post war with 230 Squadron; his recollections of operating the aircraft apply to the Sunderland V but the general principles of operation were the same for all Marks:

> Having been trained on maritime Lancasters the first difference I noticed about the handling of the Sunderland was its more sluggish control response; other than that it was a pretty sound and conventional aircraft when airborne. The big difference was its ground, or rather water, operation. We had to learn seamanship, such things as tides, currents, the rules of the water, and so on. Having arrived at the aircraft the first task was to carry out pre-start checks as with any aircraft. When it came time to move away from the mooring we started the outboard engines and had the crewmen in the bow slip the mooring line; the bollard was then stowed and the front turret moved back into position. The inner engines were then started and we manoeuvred to the take-off area. Manoeuvring the Sunderland on the water could be tricky, especially if the tides and winds were contrary, as there were no brakes – you used the outboard engines for manoeuvring and could, if required, deploy the drogues. On take-off the aircraft handled well in even a reasonably rough sea, albeit a lumpy ride with a good deal of spray, but was more of a handful if there was a contrary swell. With power on the nose was held up until a reasonable speed was reached and the aircraft was on the step, this broke the surface tension and the aircraft was soon airborne. Alighting was always tricky in certain wind and swell conditions but generally in daylight there were few problems. It was harder in a calm sea and at night when height reference was trickier and so we set a gentle descent. One of the biggest dangers was letting a wing dip as if a float hit the water; it could cause all manner of problems, most commonly that of the float fouling the aileron. The floats were only meant to be on the water for low speed manoeuvring. However, one of the trickiest manoeuvres was that of mooring up. This could take up to six crew: the pilot controlling the aircraft, a mooring party of two in the bows – the turret having been winched in – a lookout sitting on top of the aircraft to keep a watch for any problems and inform the pilot and two men in the galley to deploy the drogues. The aircraft approached the buoy at the slowest speed, using the outboard engines, to give the bow men a chance to loop the mooring rope to the buoy. The record for missing the buoy at Pembroke Dock was thirty-six attempts! My worst case was three attempts.

Ken Robinson described the particular problems of operating the Sunderland at night:

> Night take-offs and landings in Sunderlands called for special procedures and precautions. We were not able to have conspicuous flarepaths to assist us in seeing the alighting area clearly. We had to operate with only three flares, equidistantly positioned along what was considered to be a safe stretch of water. This was regularly swept by a launch of the maritime section, to remove any flotsam which might be a hazard to the aircraft. The drill on take-off was to start just before the first flare and try to get airborne by the second. If we passed the second without coming unstuck, we had to make an instant judgement as to whether the boat would get off by the third. If we were unsure of this, it was imperative that we throttled back, returned behind the first flare and tried again.
>
> When landing, we aimed to put the aircraft down between the first and second flares, so giving us the distance between the second and third to come off the step and slow down. The flares were really provided to mark out the alighting area, the illumination from them was of limited value for judging our precise height above the water. We therefore had to adopt a blind flying technique, which meant a powered approach. After gliding down to about 300ft, we partly opened the throttles on the outer engines and began to raise the nose, applying power from the inner engines until the rate of descent was about 200ft a minute. The aim was to put the aircraft into a normal landing attitude at a controlled rate of descent.

emergency, all serviceable aircraft to be kept on short slip with crew and one pilot on board from dawn to dusk in order to disperse in case of air raid' (204 Squadron ORB). The need to provide air cover for shipping in the Atlantic was to tax the minds of Coastal Command and the Admiralty during the first two years of the war, especially when shipping losses to U-boats began a dramatic rise – what would subsequently be known by the U-boat commanders as the 'Happy Time'. One way of providing cover for the Iceland–Faroes Gap was to station aircraft in Iceland and the RAF deployed a number of units to Reykjavik. On 14 June Wg Cdr Davis, OC 204 Squadron, took N9044 to Reykjavik for the dual purpose of delivering a fitter to repair a Supermarine Walrus and also to reconnoitre the coast to Akureyre for landing grounds and piers/docks. Two days later the aircraft was ordered to hunt for a submarine reported to be outside the harbour, HMS *Emerald* and MV *Empress of Austria* having just arrived with Canadian troops. At 1820 the port amidships gunner called 'sub on port bow' and the Sunderland dropped a smoke marker. The same procedure was carried out ten minutes later but no periscope was seen and the aircraft was called back to Reykjavik to escort the cruiser HMS *Enterprise*. This was the first Sunderland operation out of Iceland. The Sunderland carried out three more days of reconnaissance before returning to the UK, although Sunderlands operated on an occasional basis from Iceland during summer 1940. However, the following spring, 204 Squadron was to move to Iceland as Coastal Command and the Admiralty revised the strategy for the Atlantic War.

No. 204 Squadron had acquired Sunderlands in June 1939, after it been a flying boat squadron for ten years, and spent much of the first two years of the war operating from Scotland and Iceland. Peter Green Collection

The geographic scope of the war increased with Italy's declaration of war on 10 June, operations commencing in the Mediterranean and Western Desert. The same day, 228 Squadron sent three aircraft from Pembroke Dock to Alexandria (*see* Chapter 5).

Battle of the Atlantic

With the fall of Norway in May and France in June 1940 the overall maritime war picture changed. Germany now had access to naval and air facilities that gave her a vastly improved strategic position in respect of British shipping lanes. A number of steps were taken in an attempt to counter this. Shipping routes were changed where possible and aircraft were stationed at new airfields on the western side of the UK, in particular in Scotland and Northern Ireland, plus, with effect from May, Iceland. Flying boats such as the Sunderland were only one part of this redeployment but with their range and endurance they were to prove a vital element in what was soon to be a life and death struggle. Although Coastal Command had five squadrons available in Home Waters, this usually equated to only thirty or so available aircraft each day. Because they lacked an effective anti-submarine weapon their actual success rate was low, but their presence did provide reassurance to British shipping and a deterrent to the enemy.

Although U-boats were the main threat to Allied shipping, and the one that most of Coastal Command's maritime patrol aircraft were used to counter, there was also a significant threat posed by powerful German battlecruisers. A Sunderland found the *Scharnhorst* on 21 June, steaming south at 25kt some eight miles west of the Utyoer lighthouse. The aircraft shadowed

No. 228 Squadron was in Alexandria, Egypt, when the war broke out, but within days had moved back to the UK. Peter Green Collection

it for an hour, during which time it was subjected to intense fire, and five Swordfish made a torpedo attack (losing one aircraft). The aircraft was then attacked by four Bf 109s, the combat lasting thirty minutes and resulting in the Sunderland being damaged but also all four Bf 109s being damaged, one claimed as destroyed. The 204 Squadron ORB for June 1940 describes the incident thus:

> Flt Lt Phillips (N9028) took off at 1227 escorted by three Blenheims. A Do 18 flying boat was sighted and was attacked by the Blenheims to no effect; gunner was killed in a Blenheim and all three returned to base, abandoning the Sunderland. Whilst circling to await the Blenheims, five Swordfish were sighted by the Sunderland. Continuing alone, aircraft altered course and at 1435 sighted an unidentified submarine, submerging fast. At 1445 sighted *Scharnhorst* accompanied by seven destroyers, position reported to base. At 1510 observed five Swordfish carry out a torpedo attack and one was seen to crash into the sea. At 1610 it was observed that the aircraft was being shadowed by an He 60, which dropped a bomb falling about 50yd away. Between 1520 and 1620 aircraft was under heavy and continuous A/A fire. At 1625 first three and then one more Me 109 appeared and carried out attacks on both quarters. At 1640 one of these was shot into the sea in flames. Meanwhile, in the Sunderland, the rear turret had been put out of action very early in the engagement and tanks were punctured. Engagement ceased and the aircraft returned to base.

VIP transport was another task entrusted to Sunderlands from time to time, most of these missions being uneventful; however, when Sunderland 'G' of 10 Squadron RAAF took Mr Duff Cooper and General Lord Gort on a mission to French Morocco on 25 June 1940 it was somewhat more 'interesting':

> The flying boat reached Rabat at seven that evening. The landing was made in difficult circumstances, for the river on which the Sunderland had to alight was not more than 150ft wide. Immediately on touching down the pilot had to use his rudder in order to round a bend of the river. A number of French Air Force officers took the pilot and his passenger to the Customs wharf, whence they went to the British Consulate. The pilot returned to the aircraft, where he encountered the Harbour Master, who informed him that he must shift the aircraft upstream lest it should be in the way of incoming shipping. This was done, and shortly afterwards a secret signal was received for Lord Gort. The pilot attempted to go ashore with it in one of the dinghies, but a police boat refused to allow him to do so. The pilot was, however, determined to get the message through and made another attempt, which was once more frustrated by the police in their boat.
>
> By now it was dark. The captain ordered all lights to be switched on. This would break the strict black-out regulations and would bring the police boat quickly to the scene. On drawing alongside, the pilot and second pilot of the Sunderland jumped on board and forced the police, under the menace of their revolvers, to put them ashore. Protesting, they did so, but stated firmly that the flying boat would be placed under armed guard unless the British officers returned to it immediately. The captain of the flying boat explained in broken French that his companion, the second pilot, was in reality the captain and that he himself had only come ashore to obtain provisions for the crew. The French police were duly deceived and allowed the real captain to go into the town. The message was delivered. A few hours later the whole party returned to the flying boat and for the rest of the night the crew remained on watch beside their guns. The Sunderland took off just before dawn and landed its passengers safely at Gibraltar, going thence to England. (*Coastal Command*)

Convoy patrols were very much the order of the day for the Sunderland units, aircraft providing cover over and around the main routes into the UK, the SW of Ireland being one of the most frequent areas:

> To keep an unblinking and vigilant look-out from the turrets and side windows of a Sunderland over what seems an illimitable stretch of sea demands physical and mental endurance of a high order. Sometimes a fishing vessel, British, Spanish, French, Norwegian, Icelandic, is seen; sometimes a raft, more rarely a periscope with a spume of foam about it. When that is sighted or when the submarine is seen on the surface, the klaxon sounds and the crew get ready for immediate action. (*Coastal Command*)

A number of aircraft losses were never explained, the Sunderland simply failing to return from a patrol with no word received by radio; 210 Squadron lost one of its most experienced captains when Flt Lt Ainslee failed to return from one such convoy patrol on 29 June in N9026.

The second Sunderland U-boat kill came on 1 July 1940, Flt Lt Gibson of 10 Squadron RAAF (P9603) attacking and sinking U-26, the submarine having recently torpedoed the SS *Zarian*. Diving to attack, the Sunderland dropped a salvo of bombs when the U-boat was at periscope depth. It then surfaced and so a further four bombs were dropped. The attack was accurate and the crew were seen to jump into the water as the submarine began to sink. This was a Type IA boat of 862 tons and was on its sixth war cruise, having left on 20 June for south-west Ireland.

One of 10 Squadron's aircraft (P9602) had been damaged by enemy action on 29 July, the Squadron record for 16 August providing an interesting account of the repairs that were carried out:

> Three side skin plates were torn longitudinally, with accompanying second frame, but the main damage consisted of holes amounting to nearly half the total cross-sectional area of the front starboard main frame. Repairs were carried out by personnel of the Squadron, the affected skin sheets were removed and the damaged members trimmed up by hand drilling and filling. Packing pieces were then fitted. Strap plates, of area equivalent to that removed, were bolted and riveted along the main starboard frame. All parts were suitably protectfully (*sic*) coated. The appropriate skins were then made up and replaced, approximately 2,200 rivets had to be drilled and replaced. (10 Squadron RAAF, August 1940)

The first depth charge attack on a definite U-boat target was made on 31 July, although without apparent result. In August, 10 Squadron had a detachment of two aircraft in Gibraltar; on 2 August both aircraft were tasked to co-operate with HMS *Argus* and its reinforcement flight of Hurricanes (and two Blackburn Skuas) for Malta, the Sunderlands providing cover in case any of the fighters had to ditch. They also carried twenty-three airmen to Malta to act as a servicing echelon for the fighters, this latter part of the task was accomplished but the flying boats failed to find the aircraft carrier. The aircraft then flew VIP passenger trips for a few days, returning to Mount Batten on 7 August.

The majority of these convoy sorties were notable only for hours spent over the cold, dark sea, eyes searching for a glimpse of an enemy submarine, or perhaps the survivors from a torpedoed ship. There were, however, occasional encounters with U-boats. On 16 August, Fg Off Baker (210 Squadron, P9624) was airborne from Oban on convoy patrol; during the patrol the crew spotted a conning tower and in the first attack dropped a single depth charge, which landed about 20ft ahead of the conning tower. After a second depth charge attack,

> ... the submarine was blown right out of the water, heeled over and sinking sideways. Four anti-submarine bombs were dropped 30yd ahead of the wash, a 30ft diameter gush of oil was followed by a large air bubble. The convoy SNO signalled 'Well done, hope you get your reward.'

The crew was credited with damaging U-51 – the first such success using depth charges. During the same sortie the Sunderland reported the loss of the *Empress Merchant*, and Sunderland L2163 (Flt Lt Frame) located three lifeboats, although in an attempt to alight he damaged the starboard float and had to take-off again without effecting a rescue.

The results of this attack caused much debate; an appendix to a letter submitted by Captain A/S stated:

> In view of the immediate success attending the attack on a U-boat by a Sunderland aircraft using depth charges as opposed to bombs, it is submitted that it is extremely urgent for all aircraft operating in U-boat-infested areas to be fitted for carrying as many depth charges as possible.

A few weeks later the Air Ministry wrote to Coastal Command:

> The following is recommended for depth charge attack using Mk VII Depth Charges – a stick of three in each attack with spacing not less than 50ft and set at depth of 100ft unless local conditions suggest otherwise, height not more than 150ft and speed not more than 150kt.

Towards the end of the month, on 29 August, Plt Off Baker of 210 Squadron attacked a U-boat in conjunction with a number of destroyers, the latter claiming that the submarine had been destroyed, although no credit was subsequently given.

Unusual sorties were called for from time to time; one such was flown on 4 September by Sqn Ldr Gibson (P9603) of 10 Squadron RAAF:

> ... ordered to search for three lifeboats which contained survivors of a Norwegian MV, *Tropic Sea*. This merchantman had been intercepted in the Indian Ocean and was being taken to Europe by a German prize crew and was scuttled when the British submarine HMS *Truant* intercepted the vessel on 3 September. The British prisoners of war were taken on board the submarine and the object of the flight was to pick up the German members of the crew and bring them to England for interrogation. As it was anticipated that the Germans would offer some resistance, all members of the crew were armed with revolvers. At 1613 the three lifeboats were seen ... the aircraft circled ... the boats then tried to group themselves together ... as no attempt was made to lower the sails, the front gunner was ordered to fire a short burst ahead of the leading boat. They immediately lowered their sails and the crew ceased to row. The aircraft was waterborne at 1520 hours and on completion of the landing run was 60yd from the smallest boat. The aircraft then slowly circled the boats and the Captain indicated that he wanted the smallest boat to come alongside.

Sqn Ldr Gibson questioned the crewmen to discover if any were German, having been told 'yes' he took them aboard and placed them under guard in the wardroom; however, when he subsequently questioned them they all claimed to be Norwegian and said that the twenty-seven Germans were in one of the other boats. A second landing was too hazardous and so the aircraft flew back to Mount Batten. Two aircraft carried out a search the next day for the other two lifeboats but with no luck.

A Sunderland located the crew of the SS *Stangrant* on 14 October, the ship having been sunk two days previously. The aircraft dropped a container of food and cigarettes to the twenty-one men in the lifeboat but was unable to land and pick them up due to the sea conditions. Two days later the Sunderland set out again:

> It was still dark when one of my gunners reported a red light on the sea some miles away ... soon we could see the outline of a boat below us. We flew around for a quarter of an hour waiting for daylight ... I discussed landing with my co-pilots. We decided that it could be done and I came down on what appeared to be flattest area of sea in the neighbourhood.

The crew were picked up, and treated to a hot breakfast on the way back to land.

A few days later it was the turn of a Sunderland crew to be rescued by a boat. Sunderland P9620 of 204 Squadron had departed at 1700 (October 29) on a special mission:

> Two hours later a magnetic storm of the first magnitude developed. This put the wireless set partly out of action and gravely affected the compass. After seven and a half hours the Sunderland succeeded in making a signal saying that it was returning to base. It received none of the replies in return. Five hours later an SOS followed by a request for bearings was picked up at base and Group HQ. By then it was six in the morning but still dark. The Sunderland, its compass unserviceable, was lost and had no fuel left. The captain decided to alight.
>
> The gale was now blowing at 80mph and the navigator judged the waves to be more than 20ft high. Three flame-floats were dropped, but they did not burn, and the direction of the wind was gauged by a parachute flare. The captain brought the flying boat down in the trough between two waves. It was lifted up by one of them, so large and powerful that it took all flying speed away from the boat, which came to a halt with both wing-tip floats intact. The crew were at once prostrated by violent sea-sickness and this endured for many hours. The wireless operator began to send out signals, not knowing if any would be received. One was, and they presently picked up a message telling them that a warship would arrive in eight hours.
>
> The Sunderland began to drift amid tumultuous seas at a speed of about 8mph. How long she would endure the buffeting was hard to say. The wireless set was dismantled, repaired and re-assembled. The signals subsequently made were picked up by the warship, faint at first, but strong after midday: 'Hurry, cracking up.' Fifteen minutes later she was sighted and the look-out on the bridge of the warship read the word 'hurry' flashed by a lamp. At that moment as the crew caught sight of the warship a wave larger than the rest caught the Sunderland head on. She began to break up and the crew – there were thirteen of them – were flung into the water.
>
> The captain of the warship manoeuvred her so as to approach the wreckage of the flying boat from the lee quarter. He took the way off his ship as the crew swept past abreast of, and almost as high as, his bridge. A naval commander and twelve ratings with lines went over the side and pulled on board nine of the crew, who had then been fifty minutes in the sea. The other four were lost. The Sunderland had remained afloat in a full gale for not quite nine hours.

Sqn Ldr Cummings (T9045) was tasked to carry out an airborne search for the missing aircraft:

> The weather was very bad in the area and a gale was blowing. Whilst on the search they heard that the crew had been rescued by HMAS *Australia* and therefore decided to return to base, their ETA being 1950. The weather was dark, with low cloud and rain. They were unable to get bearings from Inverness for Sullom Voe and

decided to make for Invergordon, the new ETA being 2020. At 2020 the aircraft was still over water; at 2110 an airfield was sighted but because of the bad weather they could not identify it although they thought it might be Lossiemouth; the aircraft was fired on. They set course for Invergordon, ETA 2136. There was no sign of base at 2140 and as lights were seen on the water they decided to land. This was accomplished as near as could be judged head on to wind, but owing to darkness it was not observed that the wind angled across the swell. With about twenty minutes of petrol remaining, a landing was made, during which the starboard float was smashed. A trawler in the vicinity came alongside and lines were passed. Attempts to tow the aircraft were foiled by the roughness of the sea and after the port float had been stoved in by the trawler, it was decided to abandon aircraft. All personnel were safely transferred to trawler A76 and the Sunderland sank. (204 Squadron ORB).

In a letter of 8 November to HQ Coastal Command, OC 204 Squadron put forward a very radical concept for future flying boat operations:

That a re-organization of administrative arrangements would lead to greater operational flexibility in the use of long-range flying boats. At the moment these aircraft are hampered by:

a. Having to return to their base, at least at intervals, for maintenance purposes.
b. Being unnecessarily restricted by having too few bases at which night or bad weather landings may be made.

I would suggest, therefore, that the operational unit of the Squadron should be abolished for long-range flying boats. Instead, Stations fully equipped for night and bad visibility landings would be established at points chosen for operational and meteorological considerations. Each of these stations would have a full meteorological and Homing D/F service, and admin. staff. A suitable workshop and small maintenance staff, a Marine section and Defence personnel would complete the establishment.

Each flying boat would then operate as a self-contained unit. It would be possible to operate in weather which is now out of the question and a suitable arrangement of bases would enable patrols to be carried out which now necessitate makeshift devices.

There were very strong points in favour of this suggestion, but other than an acknowledgement from Command there appears to have been no real debate, and nothing more is recorded.

Mount Batten was subjected to a number of attacks during the latter part of 1940, heavy raids being aimed at Plymouth. The records of 10 Squadron RAAF detail the results of many of these attacks, that of 27 November causing the loss of two aircraft:

The oil bomb struck a direct hit on one of the Squadron hangars. Aircraft N9048, which had been beached late in the afternoon, caught fire immediately and was completely destroyed. Explosions severely damaged six Pegasus engines and other equipment. The hangar was completely wrecked by the fire. ... A stick of bombs caused a fire on P9601, which was at his moorings. The port wingtip caught fire at first and the fire then spread to the front of the fuselage and along the starboard wing. Two airmen on the boat were slightly injured. Two other aircraft moored in the Cattewater (P9605 and T9047) were slightly damaged by shrapnel.

The depth charge had proved its value as an air-dropped anti-submarine weapon and Coastal Command was keen to have its aircraft carry the maximum number of effective weapons; however, as the Station Command at Pembroke Dock was forced to point out on 30 December,

Tactical Instruction No. 18 states the load to be eight Mk VIII DCs – it is pointed out that with eight Mk VIIIs aboard the Sunderland II it is not possible to close the bomb doors as the forward and central roller-catches on the doors strike against the outboard depth charges. It is suggested therefore that six Mk VIIIs DCs plus two 250lb A/S bombs be carried.

A memo was sent to all groups a few days later outlining the change – pending modification to the doors.

Whilst U-boats were the greatest threat to convoys, German long-range aircraft such as the Fw 200 Condor acted as reconnaissance aircraft to help locate convoys and then guide the hunting U-boats or surface warships to their targets; these aircraft were, therefore, prime targets, though not easy to tackle. Flt Lt Lindsay and crew in a 201 Squadron Sunderland engaged a Condor in the Iceland–Faroes gap. Ray Heard was the navigator on board the Sunderland:

We were ordered to action stations; I took up my position in the front turret, clamped a magazine on to the single VGO gun, switched on the reflector sight and waited, with some trepidation, for the action to begin. Someone in the astrodome was giving range estimates as the Condor turned towards us for a pass along the port side. We dropped almost to sea level and the Condor continued closing. By the time its silhouette more than filled my gunsight I opened fire, anticipating the controller's order by a second or two. While I could see my tracers arcing towards the target, I did not see any hits, and at our approach speeds the entire action was over in seconds. I could hear the two dorsal guns blasting away but the surprising thing to all of us was the lack of return fire from the Condor which banked hard away from us and escaped. At this point the captain sent a chill through us all, shouting 'the port tanks, the port tanks!' We all thought they had been hit but it turned out that someone had forgotten to switch over fuel tanks and one engine was beginning to cough; and all this close to the sea in pursuit of the Condor. (From *On the Step, a history of 201 Squadron* by Jeremy Nash, Moravian Press, 1990)

It later transpired that the Condor had indeed been hit and had crashed in the Bay of Biscay while trying to make it back to its base. No. 210 Squadron also had an encounter with a Condor in January, Fg Off Aikman (L2163) being attacked on 29 January:

The Sunderland held fire until the fire control officer, Plt Off Larbalestier, gave the word, bursts seen to enter cockpit and both port engines, the port inner reduced to windmill and the Fw 200 broke away. The Sunderland was hit in the rear turret and the rear gunner injured in the legs. (210 Squadron ORB, January 1941)

As January 1941 arrived the Sunderland squadrons were still primarily tasked with convoy patrols and the occasional anti-submarine search. Encounters with U-boats were still rare, although 210 Squadron put in two attacks during the first week of the month, Flt Lt Van der Kiste dropping depth charges on a suspected periscope on the 5th, damaging the bottom of the Sunderland with shrapnel. When the flying boat landed at Bowmore it began to rapidly fill with water and heavy pumping and baling was needed until a diver could apply a leak stopper from the outside. (The aircraft was flown to Pembroke Dock the following day for repairs.) The second attack was made on 6 January by Flt Lt Baker, again without result.

Mount Batten was a dangerous place as German bombers carried out attacks on the city of Plymouth; on the night of 27 November 1940 the 10 Squadron hangar was hit and Sunderland N9048 destroyed. Ken Delve Collection

On 15 January 210 Squadron received a warning order to detach three aircraft, with crews, maintenance personnel and spares, to Freetown, Sierra Leone for operations; P9623, L2163 and T9041 were duly prepared and departed Oban the following day for Pembroke Dock. The same day, the detached flight was declared to be 95 Squadron and the aircraft, with N9027 taking the place of L2163 as it had better engine hours, were given an additional 355gal (1,600ltr) petrol tank in the bomb compartment, bringing the total to 2,400gal (11,000ltr). The ground party left Glasgow on the SS *Highland Brigade*.

On 21 January 1941 Sunderland T9049 of 201 Squadron (Sqn Ldr Cecil-Wright) was on patrol along the Norwegian coast between Trondheim and Narvik when the crew spotted a German parade at a town near Narvik. They dropped a general purpose bomb on this tempting target and then used the rest of their bomb load on a nearby barracks, a motor convoy and a large ship in the harbour at Narvik:

> Immediately afterwards the Sunderland was hit by two bursts of AA fire, the first putting both front and rear turrets out of action, the second damaging the tailplane. On the way home, as the flying boat was nearing Scotland, the clouds closed right down. A landing was made on the sea near an island and the boat had to be taxied up and down in the lee of a cliff for the whole night, thirty-one vain attempts being made to get the anchor to hold. The Sunderland was towed at dawn to a nearby cove and beached. Its subsequent adventures included a duel with a Me 110, which it beat off, though armed only with a borrowed Tommy gun, when on passage to the South of England for repairs. (*Coastal Command*)

This latter sortie was flown by Flt Lt Raban, the aircraft landing safely at Felixstowe.

The U-boat war escalated during the early part of 1941, the number of sightings and attacks reflecting this – although success was to evade the Sunderland squadrons throughout 1941. Typical of the attacks on U-boats was the 9 March account by Sqn Ldr Birch (T9047) of 10 Squadron RAAF:

> As it was estimated that the aircraft would only be able to escort the convoy for approximately forty minutes, the two port depth charges were hauled inboard in order to reduce fuel consumption. At 1153 hours a submarine was sighted on the surface two miles away. The aircraft immediately altered course and attacked with two depth charges from a height of 30ft. At the instant the depth charges were released the tip of the U-boat's periscope was still visible. The first depth charge exploded 20ft on the starboard side and level with the bow of the U-boat; the second exploded 30ft ahead of the bow and 20ft on the port side. A smoke float was dropped 200yd along the U-boat's track and a second attack was made at 1155 with the port depth charges. No results were observed and a submarine marker was dropped. (10 Squadron RAAF ORB, March 1941)

The Sunderland homed two destroyers to the area and they put down patterns of depth charges, still without apparent result.

In-Flight Catering!

A typical menu for a long maritime patrol sortie might be:

Breakfast	Cereal, bacon and sausage, tea, bread and butter.
Lunch	Soup, half the quantity of steak carried cubed and stewed, potatoes and vegetables, dried fruit, orange.
Tea	Poached or scrambled egg, bread and butter, tea.
Supper	Remainder of steak fried, potatoes and vegetables, bread and butter with cheese.

Between the four hot meals chocolate and barley sugar were eaten, and cocoa, tea or other hot drinks were provided by the cook, so that the crew could eat or drink something every two hours.

> The Sunderland had a useful galley with a two-ring pressure stove, racks for stowage of crockery and stowage for food. Our main wartime flying rations were tinned Maconachy's stew. This was disliked by many but I actually enjoyed it. Members of the crew turned out to be very acceptable cooks. I remember that one of their 'specials' was sausages (soya!) and corned beef fried in thick batter – delicious! Tea, coffee or cocoa were regularly available and helped to keep us active during the long, lonely patrols. (Dundas Bednal, *ibid*)

Catalinas began to arrive at 210 Squadron in April and by May the Sunderlands had gone, mainly to other units. Meanwhile, from early April 204 Squadron had most of its strength operating from Reykjavik, using the SS *Manela* as a depot ship. One of the earliest events noted in the Squadron diary for this period was the loss of an aircraft: on 23 April, Sunderland N9023 (Flt Lt J. Hughes) crashed into high ground when getting airborne for a convoy escort task, three of the crew being killed.

Pembroke Dock in May 1941, with L5802 wearing 95 Squadron codes.
Peter Green Collection

Hunt the *Bismarck*

The Iceland-based Sunderlands were soon involved in one of the most dramatic naval encounters of the early part of the war. On 19 May 1941 the German battleship *Bismarck* left Kiel on her maiden voyage, in company with the cruiser *Prinz Eugen*, as Germany attempted to put even greater pressure on the vital British transatlantic re-supply routes. Two days later the warships were anchored in Dobric Fjord, Norway prior to heading for the Atlantic. They were spotted by an RAF reconnaissance aircraft and the hunt was on – the Admiralty knew the dangers that these ships posed and every effort was put into their cornering and destruction.

After an unsuccessful attack by Whitleys and Hudsons, the weather deteriorated and the German vessels were lost. Part of the major air effort in trying to locate the ships included Sunderlands and Hudsons to reconnoitre the Faroes–Iceland gap, the Sunderlands maintaining a patrol from 0615 to 2115 and covering 2,000 miles in a single sortie, 'but the weather was against them. They encountered strong headwinds, fog, rain-squalls and heavy cloud in which severe icing conditions developed.' (*Coastal Command*). The German warships were spotted in the early evening of 23 May by Royal Navy ships in the Denmark Straight, and a Sunderland from Iceland set off,

Sunderland II

One of the major changes with the Sunderland II was the use of Pegasus XVIIIs with two-speed superchargers. Another significant change was with the armament of the aircraft, an FN7 twin 0.303in turret being put in the dorsal position instead of the previous, somewhat exposed, Vickers K-gun mountings. The tail turret was also changed, the FN13 giving way to the FN4A, the latter having a greater number of rounds (1,000) per gun.

Most Sunderland IIs were also given ASV II, distinguished by four vertical dipoles and sixteen transmission loops. Only forty-three Sunderland IIs were built, twenty of these by Blackburns at Dumbarton, the remainder by Shorts.

MAP Specification dated 26 August 1942:

Powerplant:	Four Pegasus XVIII
Fuel/range:	1,962gal with maximum bombs, range 2,070 miles (225 miles less if Leigh Light fitted); 2,034gal with permanent tanks full, range 2,310 miles
Performance:	Service ceiling 12,600ft at max weight of 56,000lb; 16,100ft at mean weight of 48,500lb; max speed 194mph (211mph if depth charges internal), most economical speed 144mph; climb 10.7 minutes to 5,000ft
Armament:	One 0.303in in nose turret, 700 rounds per gun four 0.303in in tail turret, 1,000 rounds per gun two 0.303 in top turret, 1,000 rounds per gun 8 × 250lb external, or 4 × 500lb external, or 4 × 475lb DC

> … in the long twilight of those far northern latitudes to search for the enemy. … Meanwhile the Sunderland (L5798, Flt Lt Vaughan, 201 Squadron) had arrived in the neighbourhood of HMS *Suffolk* and, on sighting this ship, saw at the same time a flash of gunfire well ahead. 'As we closed', says the Captain in his report, 'two columns, each of two ships in line ahead, were seen to be steering on parallel courses at an estimated range of 12 miles between the columns. Heavy gunfire was being exchanged and the leading ship of the port column was on fire in two places, one fire being at the base of the bridge superstructure and the other farther aft. In spite of these large conflagrations she appeared to be firing at least one turret forward and one aft.' At first the captain of the Sunderland could not identify the burning ship. He turned towards the starboard column and noticed that the second of the two ships composing it was making a considerable amount of smoke, and that oil escaping from her was leaving a broad track upon the surface of the sea. He approached nearer, and as he did so the ship on fire in the column to port blew up [this was HMS *Hood*].
>
> A few seconds later the Sunderland came under heavy AA fire at the moment when its captain was identifying the ships in the starboard column as the *Bismarck* and *Prinz Eugen*. He was

forced to take immediate cloud cover and when, five minutes later, he emerged into an open patch, the ship which had blown up, and which he now realized was British, had almost completely disappeared. When the Sunderland flew over the spot all that could be seen was an empty raft, painted red, surrounded by wreckage in the midst of a large patch of oil. (*Coastal Command*)

Despite this success against the *Hood*, and damage caused to her companion, HMS *Prince of Wales*, *Bismarck* was doomed; she parted company with *Prinz Eugen* and made towards France. The weather was still poor and the shadowing British ships kept losing contact; air reconnaissance was now the key and although none of the Sunderland patrols played any further part, the RAF's Catalinas were involved. The honours, however, really belonged to the Fleet Air Arm Fulmar and Swordfish crews (from HMS *Victorious*): in a series of daring torpedo attacks they caused crucial damage to the battleship. On 27 May the British naval force closed in and sank the *Bismarck*.

Defending the Boat

Deliveries of the Sunderland Mk II to 201 Squadron commenced in mid-1941 and another shift around of units saw 204 Squadron leave Iceland for Gibraltar in mid-July, but not before a dramatic incident, on 10 June, in which Sunderland N9047 caught fire and sank at its moorings:

> After alighting at Skerjafjordur (Reykjavik) aircraft N9047 was refuelled. About an hour afterwards the signaller in *Manela* reported that smoke was issuing from 9047 at her moorings. Three minutes later he reported her on fire and flames with black smoke billowing from above the mainplane. Within two and a half minutes the mainplane had broken and tipped over; at 1730 the aircraft was two-thirds under water, blazing furiously and sinking fast. Those of the crew still aboard had been got off by motor launch and nobody was hurt. Overnight salvage work was begun and by morning the four motors had been recovered. The fire had not reached them – much gear was recovered and by the 14th the mooring had been cleared of wreckage. (204 Squadron ORB, June 1940)

In June 1941 Coastal Command issued instructions for trials with new camouflage schemes for the side and underneath of their aircraft, eggshell blue, pale pink, light grey and matt white being evaluated. It was found that for the prevailing weather conditions in the North Atlantic, matt white for side surfaces and gloss white for under surfaces was the most efficient combination. An order to this effect was issued on 8 August, although it was spring 1942 before all aircraft were completed.

Amongst the 'good advice' transmitted via the pages of the *Coastal Command Review* were the following two items:

> Sea-gulls and gannets are found to be extremely useful targets for gunnery practice. Their constant changes of course make them difficult to hit; their airspeed, though low, is enough to necessitate deflection; and yet they are passed so quickly that rapid thinking and snap shooting are essential. The gunner who brings one down is rewarded by 5 shillings when his claim is confirmed by another gunner; it is rare for anyone to earn it even twice. The drill is that a first sighting report comes from the fire control position, 'one sea-gull, 100yd, port bow, shoot him down'.

The second item referred to navigation training for rear gunners:

> The construction of the Sunderland makes it impossible to take an accurate drift reading from the front on a smoke or flame float. By courtesy of Messrs Frazer-Nash, an excellent drift sight has been manufactured for a four-gun turret; it is relatively simple to graduate the Rear Gunner's inner turret ring and the four guns which serve instead of drift wires. In flying boats with this equipment the Rear Gunner is given some instruction in navigation so that he appreciates the importance of obtaining accurate drifts at frequent intervals, a process which had the additional advantage of giving him an interest to relieve the tedium of his isolation. A definite note of superiority can be detected in his voice as he reports to the Navigator 'a drift of 4½ degrees to starboard'. His activities are specially valuable at night when they save the Navigator a series of hurried walks to the rear turret to observe flame floats.

On 30 June 1941 Sunderland P9600 (Fg Off A. Wearne) of 10 Squadron RAAF was involved in a combat with an Fw 200, the aircraft suffering damage:

> The port midship gunner reported that the port outer engine was losing large quantities of oil. LAC Griffin, the first Fitter of the aircraft, volunteered to crawl out into the wing to see what could be done. After an investigation he reported that there were two large holes in the bottom of the side tank, that nearly all the oil had gone and that the nacelle was filled with it. He then obtained some plugs and went out to the port outer engine and stopped the leaks. He then returned to the aircraft to procure a funnel, two gallons of oil and some tools in order to pierce the top of the tank. He did this and put the spare oil into the tank by using an old peach tin as a ladle. This work was done in a very cramped space and in intense heat. This airman crawled into the wing for the fourth time and ladled a further two gallons of oil into the tank. The whole operation took approximately two hours. LAC Griffin's action was probably responsible for saving the engine and considerably increased the safety of the return journey. (10 Squadron ORB, June 1941)

LAC Griffin was awarded the DFM for his action. Repairing oil leaks in flight seemed to become a 10 Squadron speciality and similar adventures are recorded a number of times, the next occasion being 24 October, Flt Lt Wearne's aircraft (P9605) having been hit by fire from an 'unknown vessel'.

> After an investigation the second Fitter (LAC Hunter) reported that there were two large holes in the port outer tank. The first and second Fitters (AC1 King) then crawled out to the port engine with some tools and plugged the holes. The second Fitter returned to the cabin for a five gallon drum of oil and a hatchet while the first Fitter kept watch on the engine. A hole was cut in the top of the tank and the tank was refilled from the spare drum of oil. Both Fitters then returned to the cabin. While the second Fitter kept watch on the pressure and oil gauges, the first Fitter returned to the port outer engine with a jug. He remained there for the next two hours and kept the engine running smoothly by collecting the oil from the nacelle of the engine and pouring it back into the tank. (10 Squadron ORB, October 1941)

August saw 228 Squadron return to the UK, being based initially at Calshot; however, in early October facilities at Stranraer were being looked at. At last the Squadron was told on 26 October that three new aircraft had been allocated (W3989, W3990, W3991) and were to be collected from Pembroke Dock. The first of these, W3989, duly arrived at Stranraer on 3 November. Meanwhile, on 13 October the advanced party of 201 Squadron settled in to its new home at Castle Archdale on the shores of Lough Erne. To route to the Atlantic and the patrol areas, the unit's Sunderlands flew over Eire along an agreed air route, known as the Rathlin corridor.

Declining Numbers

A Coastal Command requirement issued in late 1941 stated that a minimum of 818 aircraft of all types was required in order to meet future operational requirements, this figure including 26 flying boat squadrons (equipped with 150 Catalinas and 72 Sunderlands). However, the supply position in respect of flying boats was poor, in the case of the Sunderland this not being helped by:

> ... labour troubles in Belfast and a lack of interest in the type [by Shorts] as compared with the Stirling, shown by the fact that the manufacturing firm had brought Sunderland production almost to a full stop. The new factories at Windermere and Dumbarton were not expected to deliver any boats until 1942 and only four per month were being produced at Rochester. (AHB *Narrative*, 'The U-boat war')

The company argued that it could build six Stirlings for the same effort as three Sunderlands and this was used by AOC-in-C Coastal Command, in a December memo to the Chief of Air Staff, to suggest that the company produce Stirlings for Bomber Command whilst Coastal Command could be given long-range B-24 Liberators from the American delivery pool. However, the same month a decision was made by the Air Staff to equip a further nine flying boat squadrons; the AOC responded that:

> A large increase in flying boat squadrons would create serious training difficulties. The Command possessed only one flying boat OTU which was quite incapable of providing crews and backing for the proposed nine squadrons, and apart from seventy-five aircraft required for the actual squadrons another forty-six would be needed to make good deficiencies in the present OTU and provide for a second. (AHB *Narrative*, 'The U-boat war')

By autumn 1941 a large part of the Sunderland strength had moved to West Africa Command, leaving Coastal Command with only two squadrons, 201 Squadron at Sullom Voe and 10 Squadron RAAF at Pembroke Dock; it was to be almost a year before any significant increase (other than the return of 228 Squadron) in Sunderland strength took place in the UK.

On occasions targets other than submarines received attention from the Sunderlands; W3984 of 10 Squadron RAAF was involved in one such operation on 23 December: 'At 0916 the W/T operator intercepted a signal from K/10 reporting the location of an enemy tanker.' The aircraft was ordered by Group to continue its patrol, but a little while later actually found the tanker: 'As our aircraft circled while taking photographs the tanker opened fire with heavy AA but fire was not returned as the Captain considered range too great.'

The following day Flt Lt Costello (W3984) was on an anti-shipping patrol when he came across a 12,000-ton tanker – probably the same one:

> The tanker appeared to have slight list to starboard and was slightly down at the stern, leaving an oil streak. As our aircraft manoeuvred to attack, a Ju 88, which was escorting the tanker, approached on the port beam. Our aircraft turned to starboard and opened fire with rear guns at a range of 400–600yd. Both aircraft exchanged fire and then the enemy broke off and disappeared into clouds. Flying at a height of 1700ft, an attack was made on the tanker with six 250lb depth charges and two 250lb A/S bombs, set at half-second delay. As the observer was acting as Fire Controller, the depth charges were dropped by the Pilot without the use of the bombsight. An A/S bomb, which exploded 30yd ahead of the track of the tanker, was the nearest explosion. The tanker kept up fire from two fairly large-calibre guns. A British destroyer was sighted at 0930 and signals were exchanged informing the destroyer of the position of the enemy.

The aircraft then continued to shadow the tanker, passing position information to the destroyer. The Sunderland then engaged an enemy aircraft that had bombed, and missed, the destroyer. A successful torpedo attack subsequently sank the tanker – much credit for this success being given to the Sunderland crew.

No. 202 Squadron, based at Gibraltar but under Coastal Command control, flew its first Sunderland operation on 26 December, Flt Lt Cooper taking T9084 on a routine ASP; although most of its Catalinas had now gone, the squadron was still operating the two types (and did so to 1945).

Sunderland III

The prototype Sunderland III, the converted T9042, was first flown at Rochester on 28 June 1941, one of the main changes being a faired main step to the hull. This was to be the definitive and most-produced variant of the Sunderland, some 456 aircraft being built.

The first production aircraft was W3999, its first flight taking place on 15 December 1941. This was the last major airframe change for the Sunderland, subsequent variants revolving around engine and equipment changes.

MAP Specification dated 30 August 1942

Powerplant:	Four Pegasus XVIII
Fuel/range	1,862gal with maximum bombs, range 1,895 miles (235 miles less if Leigh Light fitted); 2,137 gal with permanent tanks full, range 2,235 miles; 2,548gal with auxiliary tanks full, but 1,200lb of equipment has to be shed, range 2,970 miles
Performance:	Service ceiling 11,700ft at max weight of 58,000lb, 15,600ft at mean weight of 49,900lb; max speed 196mph (213mph if depth charges internal), most economical speed 151mph; climb 11.8 minutes to 5,000ft
Armament:	One 0.303in in nose turret, 700 rounds per gun four 0.303in in tail turret, 1,000 rounds per gun two 0.303 in top body turret, 1,000 rounds per gun 8 × 250lb external, or 4 × 500lb external, or 4 × 475lb DC, plus same again internal fit; total max bomb load 4,000lb (3,800lb if DC)

In a second data sheet for the Sunderland III, dated 18 November 1943, there were a number of changes in respect of performance:

Operational:	2,137gal for 2,500-mile range, loiter endurance 18.2hr
Max bomb:	1,861gal for 1,968-mile range, loiter endurance 14.7hr
Max fuel:	2,548gal for 3,120-mile range, loiter endurance 23.2hr
Speed:	Max 212mph at 1,500ft; most economical, 143mph; loiter 131mph

CHAPTER THREE

Home Waters and the Atlantic

The Years of Success

This view of W3999 in April 1942 shows the boat-shaped hull (planing bottom) to good effect. Peter Green Collection

Operating from Stranraer, 228 Squadron were flying anti-submarine escort for convoys from early February 1942, the squadron having received its full complement of Sunderlands. Whenever major airframe repairs were required the work was carried out by Scottish Aviation at Greenock, typical of this being hull repairs to W3992 on 21 February and T9085 four days later.

The Navy was famed for firing at friendly aircraft – they even managed to mistake the unique bulk of the Sunderland on occasions, 201 Squadron's ZM-S suffering damage in one such incident in March 1942. On this occasion the aircraft suffered only light damage; however, on 29 May the result was more tragic when ZM-P (Fg Off J. R. Traill) was shot down by the convoy it was escorting and only one crewman, Sgt Wheatley, was rescued.

By February 1942 a total of sixty-four Sunderlands had been fitted with ASV. A new installation layout was approved in April 1942 and a trial installation was agreed in July 1942 (*see* ASV box on page 24 for more details).

A large number of excellent attacks on U-boats were made by aircraft during March and April; of the twenty-three recorded incidents, four involved Sunderlands. No. 228 Squadron moved to Oban in March; the routine of patrols and escorts soon producing results:

> March 13, Sunderland N/228, 1920 hours, flying at 300ft in haze sighted a half-submerged U-boat half a mile away, making six knots. The beam aerials were in use but gave no contact. Seven depth charges were released at 50ft at an angle of 30 degrees to the U-boat's course, 15 seconds after its disappearance. They straddled its line of advance 100yd ahead of the swirl, therefore, presumably 12yd ahead of the conning tower. Two members of the crew saw the stern above water between two of the explosion marks. The U-boat must therefore have been forced up by the explosion and was probably damaged. Position 6035N 1400W.

This attack by Flt Lt Cooper and crew was the Squadron's first recorded 'success' using radar equipment; the eighth depth charge, which had hung during the attack, was later successfully jettisoned. The official Coastal Command record of U-boat 'scores' does not given any credit for this attack.

The Australian 10 Squadron was still operating from Mount Batten. On 20 March

> R/10 was on anti-shipping patrol in position 4343N 0143W sighted a motor launch, approx 50-60ft long, with deckhouse amidships, and flying the German flag, it was then 0745. The aircraft attacked four times, but the first was unsuccessful as the depth charges hung up. The

8 May 1942 with B/10 Squadron RAAF on patrol – note the oil streak in background. Ken Delve Collection

Flt Lt Pockley and crew attack a U-boat on 28 May 1942, but with no definite result. Ken Delve Collection

second was made with machine-gun fire, many hits being scored, and pieces were seen flying off the vessel. On the third attack, four 250lb depth charges set at 25ft were dropped from 500ft. The Motor Launch was seen to be in their centre, and was blown up on the crest of the explosion, wreckage was seen when that had subsided. Three 250lb depth charges were dropped in the next attack from 500ft, with a setting of 25ft (the fourth depth charge hung up). Once again the vessel was well within the explosion area and was blown up on the crest. After this attack the launch appeared to be a mere shell floating well down by the stern, and was surrounded by wreckage. As a single-engined aircraft was seen to be approaching, the Sunderland went away. (*Coastal Command Review*)

By no means an ideal target for a depth charge – but the result appears to have been effective. Exactly a month later:

April 13, Sunderland R/201 flying at 5,000ft at 1142 sighted a wake at 12 miles' distance when in 3/10 cloud, the base of which came at 2,000ft. The aircraft approached above the clouds, but the U-boat dived when still one and a quarter miles away. Three depth charges dropped, 30 seconds after its disappearance; five more failed to release. The nearest is thought to have fallen 200ft astern. Position 5418N 1453W.

The following day another 201 Squadron Sunderland was in action:

At 1450 hours Z/201 started to search for the same U-boat [referring to a contact made by a 502 Squadron Whitley] flying at 4,000ft. The Special Equipment received a contact at 1826 hours, at 8 miles' distance to port. The U-boat was sighted from 6 miles' distance, making seven knots. It submerged when the aircraft approached. Three depth charges were dropped, four failed to release, thus enlarging the spacing to 90ft. At 1850 hours, flying at 500ft, the aircraft spotted a periscope feather, quarter mile distant, which immediately disappeared. A second attack was made with the remaining depth charges, 90 seconds later, but one again failed to release. Position 5000N 1408W.

The German view of the convoy battle at this time makes for interesting reading:

With the resumption of the convoy battle in 1942, Flag Officer U-boats soon realized that the enemy air power had grown sufficiently to provide several daily patrols of the Bay of Biscay and also additional aircraft for convoys being pursued by our boats. Our operations against OS33, SL118, SC97 and SC99 all had to be stopped prematurely owing to the sudden arrival of continuous air cover. ... The Sunderland flying boat and four-engined bombers used for this purpose were all fitted with radar and carried an adequate bomb load. The handicap to the U-boats would be correspondingly greater as soon as air cover could be extended across the whole North Atlantic.

If the U-boat was bombed while in the process of diving, she was most vulnerable; therefore, the rule was not to dive unless a depth of 60–80m could be reached before the bombs began to fall. It was always a grave decision to remain on the surface and face up to an approaching enemy aircraft, for the U-boat's only weapon was a 2cm machine-gun, which was liable to fail at the critical moment after being exposed to sea water for perhaps weeks on end. (*German Naval History – the U-boat War*)

By September 1942 there were plans to increase the anti-aircraft armament of the U-boats with pairs of 2cm guns and even 37cm quick-firing guns.

One of the major problems in achieving success against the U-boats was the unreliability of the standard depth charges, both in terms of release failures and the effectiveness of the DC itself. The introduction of the Torpex DC led to an increase in the number of attacks that caused damage to U-boats.

Flt Lt Maynard (W4017) of 228 Squadron attacked a U-boat on 28 April:

S/E contact was made but nothing was seen and the patrol was continued. The aircraft later returned to the area of the first contact and picked up an object. Homing was carried out and when within 1½ miles contact was identified as a U-boat. An immediate attack was carried out from 700ft but unfortunately five of the bombs hung up. Nos 1 and 3 straddled the U-boat and machine-guns were also used. Results were not observed. The aircraft landed at Loch Boisdale because of shortage of fuel. (228 Squadron ORB)

Sadly, Loch Boisdale was the scene of a fatal crash on 4 May, 228 Squadron's N3984 crashing on landing, with the loss of two crew.

April also saw another Sunderland squadron added to the Coastal Command order of battle, a second Australian unit, 461 Squadron, forming at Mount Batten on 25 April. A signal had been sent from Coastal Command on 14 April stating that:

> The necessity of forming this squadron as quickly as possible cannot be over-emphasized. To start the squadron it is necessary that a nucleus of aircrew and maintenance personnel be transferred from 10 Squadron. Since the OTU will be able to provide the squadron with only one crew per month it is necessary that experienced Australian aircrew personnel now in landplane squadrons be posted to 461 and converted to flying boats within the scope of the squadron. (461 Squadron ORB)

By the summer, 4 OTU was stating that it had sufficient training capacity to provide more crews. The squadron received its first aircraft, T9090, on 22 April and was also told that it would move to Poole Harbour (Hamworthy) as soon as this was available. It was indeed a quick work-up for the squadron and the first operational sortie was flown, by Wg Cdr Halliday, on 1 July, T9090 flying an anti-submarine patrol.

Sunderlands of 10 Squadron made two attacks on 8 May:

> At 0805 hours B/10, flying at 1,500ft, sighted an oil streak at least 300yd long. Two attacks were made with depth charges set to 25ft, at an interval of 26 minutes, during which the streaks changed course. In the first case the bursts were observed 200yd, and in the second 300yd, ahead across the line of the streak. After each attack the oil bubbles increased greatly. After the second, the U-boat reduced speed from three knots to one. B/10 shadowed its course until relieved by A/10, which arrived at 1305. Oil bubbles were then constantly coming to the surface at the end of the streak, and they appeared to move along, above the track. Half an hour later, A/10 attacked with four depth charges set to detonate at 25ft. At 1517 a second group of bubbles was seen rising from the edge of the oil patch near the beginning of the streak. The speed of advance was now reduced to less than three-quarters of a knot (as was ascertained by dropping a smokefloat). Four Hampdens were attracted to the scene by a second smoke float, and one of them, N/415, attacked at 1650. Five minutes later the oil appeared to thicken and spread. A/10 made a second attack at 1950 with its four remaining depth charges, set for 150ft. Half an hour later, shortage of petrol forced the aircraft to return to base. Position 4810N 0925W.

During the afternoon another squadron aircraft, 'U', had also picked up a U-boat:

> The aircraft immediately attacked in a shallow dive, the U-boat crash-dived, and had been submerged for 32 seconds when eight 250lb depth charges (set to 25ft and spaced at 45ft) were released from 50ft. The pilot estimated that the stick fell short, but that No. 8 exploded directly on the line of advance, 150yd ahead of the swirl. An oil streak was seen after the explosion in the disturbed patch of water. The stick was correct for distance ahead of the swirl but fell slightly short of the side, so that No. 8 and No. 7 exploded about 13yd abaft the conning tower, but after 37 seconds the U-boat was probably at 70 or 80ft depth, well below lethal range; still it probably got severely shaken. This is the first case in which the all-important interval between the disappearance of the U-boat and the release of the depth charges had been taken by stop watch, also marked in yd. It seems to have helped because the pilot was correct in his estimation of how far ahead to drop the stick. Position 4744N 1047W.

Flt Lt Wood (W3986) damaged U-71 on 5 June 1942. Ken Delve Collection

Sunderlands also recorded attacks in the latter part of May. Late May was a hectic period for 228 Squadron, not so much in terms of operational sorties – it was rare for more than two sorties to be mounted in a day – but rather in respect of sightings and attacks. Flt Lt Goyen was airborne at 1200 hours on 21 May in W4026:

> Carried out a sweeping line ahead search ... at 1430 sighted a conning tower of U-boat on starboard bow from height of 200ft. Dived and turned to attack but DCs failed to release. On return (from 30 miles away) at 1504, U-boat was again sighted on surface and seven DCs were dropped from 50ft. Estimated two DCs were 'near misses' and damage may have been done. Brown patch observed. Aircraft searched area for 5 hours 23 minutes before returning to base. (228 Squadron ORB)

Two days later:

> At 1133 on May 23 Sunderland N/228 flying at 4,000ft in cloud and 20 miles visibility sighted a U-boat four miles ahead making 12–14kt. It submerged at 2 miles' distance. The aircraft attacked from the starboard quarter at 75ft, and released six 250lb depth charges (torpex, 50ft setting, 33ft spacing). Two more, Nos 5 and 8, failed to release. The U-boat had already been out of sight for a minute. The first explosion took place 100yd ahead of the swirl and 75yd to the right of the line of advance. It should have been 200yd short of the conning tower. No. 7 should just have reached the line of advance but about 100yd astern. Position 6150N 0332W.
>
> On May 28 at 1350 hours, Sunderland R/10 [W3983, Fg Off Pockley, on detachment with 202 Squadron] from Gibraltar, flying at 1,000ft, sighted an Italian submarine, 5 miles ahead. The aircraft dived to attack but had to take avoiding action owing to heavy flak; the submarine carried one large and several smaller guns. A second dive resulted in an unsatisfactory run, in which, however, the submarine was raked with machine-gun fire. Another attack, made from 40ft amid heavy flak, was delivered from the port quarter while the submarine was turning to port; four torpex depth charges were released but overshot. The aircraft's port bomb circuit being unserviceable, the remaining four depth charges had to be manhandled to the starboard rack. These were dropped in another

attack made up the track from 30ft, as the submarine turned to port (they were 250lb torpex, set for 25ft, spaced to 36ft). The stick straddled the submarine, one depth charge exploding on the port side and the rest on the starboard at 30 degrees to the track. The submarine was completely hidden by spray. When this subsided, its speed had fallen to three or four knots, and the course became erratic. A large rent in the plates of the port bow threw a continual wave on the upper deck, and a large dent in the side plating near the water line, just forward of the conning tower. R/10 stayed near the submarine until a Hudson arrived at 1830, though its machine guns gave much trouble and it was repeatedly hit by flak; the starboard engine and float were damaged. Position 3759N 0208E.

The 233 Squadron Hudson also attacked with depth charges, causing further damage.

The following day it was 10 Squadron's 'Z' which was in action, having been called on to attack a submarine that had been found and attacked by a 53 Squadron Whitley:

The aircraft lost height in two turns and reduced the distance to a mile and a quarter before the U-boat submerged. Fifty seconds later, an attack was made up the track from 140ft. Seven depth charges were dropped but No. 5 failed to release. The first DC exploded 100yd ahead of the swirl; the last three DCs were right for range, and probably for line, but they are likely to have been too shallow. The U-boat must however have been very badly shaken up. Position 4628N 0957W.

Neither of these late May attacks was given any subsequent credit. However, the Italian submarine *Veniero* went missing around this time and may have fallen victim to Allied attack – perhaps by one of these Sunderlands.

The Flying Boat Training Squadron at Stranraer, equipped with a mix of Sunderlands, Catalinas, Londons, Stranraers, Lerwicks and Singapores, had been responsible for flying boat conversion training in the early part of the war. However, the growing requirement for Sunderland and Catalina crews led to this unit being expanded to become 4 (Coastal) OTU with effect from 16 March 1941. In mid-June the OTU moved to Invergordon.

In a 31 May summary, HQ Coastal Command outlined its position regarding expansion and re-equipment; in respect to flying boats this document stated that progress towards the 'Revised Target E' proposal of December 1941 – for twenty flying boat squadrons (eight with Sunderlands) each with nine aircraft, was progressing slowly and that all but one of the Sunderland units had only six and not nine aircraft.

31 May 1943: U-boat attack by DV969 of 10 Squadron and DD838 of 228 Squadron; U-563 was sunk in this attack. Ken Delve Collection

The location is unknown, but obviously a production or heavy maintenance facility; an NS-coded aircraft of 201 Squadron brings up the rear of seven Sunderlands. Peter Green Collection

June proved to be a hectic month in the war against the U-boats, with 10 Squadron RAAF in particular continuing its activities from late May. The first contact came on 5 June when Sunderland W3986/U (Flt Lt Wood) was flying over the Bay of Biscay and picked up a contact on radar:

A wake was then seen though binoculars; the aircraft dived towards it and found a U-boat making 10kt. An attack was made from 50ft at a speed of 205kt, 25 seconds by stop-watch after the U-boat had disappeared. Eight torpex depth charges were released from its starboard quarter. The centre of the stick was aimed to explode at the presumed position of the conning tower, 130yd ahead of the swirl, and is believed to have in fact exploded there; the depth charges straddled the line of advance. A minute later the U-boat surfaced, bows first, at a steep angle, an air

The growing requirement for crews led to the expansion of the Flying Boat Training Squadron into No. 4 (Coastal) OTU in early 1941; T9042 served with this unit for some while, eventually being disposed of in February 1945. Peter Green Collection

To man the increasing number of flying boats now becoming available, only one experienced pilot can be allotted to each crew. He is the captain of the aircraft, and is assisted by a second pilot posted direct to the squadron from the School of General Reconnaissance, and trained to the standard required in the squadron. After a second pilot has gained experience in the squadron, he can expect to be withdrawn and sent to No. 4 OTU, after which he will be posted to a squadron with a crew of his own. On passing out of the OTU these crews, like those from the landplane OTUs, spend the first month working up to the operational role.

Some squadron commanders doubt whether crews trained in this way will be sufficiently experienced to carry out squadron work. They therefore split up new crews, which is not only discouraging to the crews, who in many cases have learnt to think as a team, but it also prevents the Command from building up an urgently needed reserve of crews. An arduous flight from England to Aboukir via Gibraltar and West Africa was recently undertaken by one of the first crews trained in this way. The Captain's story shows what freshman crews can do:

I had been serving as second pilot in a Sunderland squadron for thirteen months, and saw no prospects of being made a captain for some months to come, when the procedure for second pilots being posted to the OTU for training as captains came into operation. My Commanding Officer would not, at first, recommend me for this training, but after repeated requests I was allowed to go.

On completion of my OTU training I was detailed, with my crew, to fly a Sunderland out to the Mediterranean. We collected it at Pembroke Dock, and after checking it and swinging the compass and loop aerial, flew it to Mount Batten. The first leg of the trip, to Gibraltar, was made by night. The navigator set about his task with calm confidence. The thick weather encountered made it difficult to get astro sights, but those he was able to obtain were accurate and the courses given brought the boat to the Straits of Gibraltar by dawn. After three long oversea flights around West Africa, the route led overland along the Congo to Lake Victoria, down the Nile Valley, and on to Aboukir. For most of the journey, the wireless operator had to contend with commercial radio organizations, including an extensive use of the Q code in French, and a dust storm code over the desert.

Two inspections became due on the way out. We had ground staff for the first time, but the '90 hour' we carried out ourselves on Lake Victoria, checking the engines, airframe, equipment and instruments. Finally we joined our squadron at Aboukir without mishap. After all the difficulties we had encountered and overcome in nearly 100 hours' flying, we had acquired a very keen crew spirit, and confidence in each other. We were therefore very disappointed when we found that it was the practice of this squadron to split up new crews amongst other crews. I was given command of another crew, and we began night operations and anti-submarine patrols. The spirit of this crew – strangers to me – did not compare with that of my previous crew, who, when separated as subordinates in other crews, soon lost the keenness and self-confidence they had acquired.

(From *Coastal Command Review* No. 4)

TA-coded DP200 of 4 (C) OTU at Invergordon. This aircraft was subsequently converted to a Mark V and was scrapped in 1957. Peter Green Collection

bubble 25ft in diameter appeared to port alongside the conning tower, and oil bubbles rose in the wake, just ahead of the explosion mark; then the bows dipped as the stern came high out of the water, and gradually the U-boat steadied, with its bows awash and listing to port. It was still moving ahead on the motors. In the next 10 minutes the aircraft poured some 2,000 rounds from the nose, port and tail guns in to the hull, upper deck and bridge structure. The U-boat moved slowly in figures of eight, finally getting under way on the Diesels. A couple of men appeared in the conning tower but vanished when the tail gun opened up on them. At 1635 some of the crew seized an opportunity, when the aircraft had ceased firing to conserve ammunition, and manned both the anti-aircraft gun, on the after end of the bridge, and the main gun forward of the conning tower. They opened fire and hit the aircraft several times but caused no casualties.

Ten minutes later the U-boat increased speed to 8–10kt. At 1731 it reduced speed and submerged. A large oily patch remained in the position of diving, and an oil streak continued to extend along the course. The aircraft sent a message to base at 1819, when the oil streak was moving at one knot, and at 1924 established wireless contact with relieving aircraft.

U/10 left at 1939, when the oil track was fading, and was immediately attacked by a Focke-Wulf. The enemy made four attacks, all from abaft, beam or astern, in the course of an hour and five minutes. In the first three, U/10 was several times hit by cannon fire and had the rear turret disabled; it was kept in action by hand. The last attack was made from close range, and the Sunderland then sustained five large and eighty small holes, the R/T aerial was shot away and both flaps were damaged. It returned fire from all guns that would bear, throttling back to reduce the range. The Focke-Wulf overshot and was hit repeatedly; it yawed at right angles, broke off the attack and disappeared eastward, flying low over the sea; probably it failed to reach its base. U/10 returned uneventfully though other enemy aircraft had come near the coast to intercept it. Some of the shot-holes were beneath the waterline, but arrangements were made to plug them as soon as it moored, and it was quickly beached. Position 4456N 0344W.

Series of three photographs showing the attack on U-71 by W3986 of 10 Squadron RAAF on 5 June 1942; the crew were credited with damaging the U-boat.
10 Squadron

The crew was credited with damaging U-71. Two days later a pair of 10 Squadron aircraft (W4019/R, Flt Lt E. Yeoman, and W3994/X, Plt Off T. Egerton) operating out of Gibraltar were given credit for causing severe damage to the Italian submarine *Luigi Torelli*:

> At 0358 hours on June 7 Sunderland R/10, flying from Gibraltar, received a contact three miles on the port beam. Visibility was 500yd, with the aid of a young moon. The aircraft circled and homed, and finally sighted a submarine making 12kt. An attack with eight 250lb depth charges was made from 100ft at an angle of 30 degrees to the track, and the centre of the stick was thought to have been about 30yd from the port beam. The submarine was then still fully surfaced and firing light flak, which hit the aircraft several times. As the aircraft broke away, the submarine opened up with heavy flak, hitting the starboard outer engine and holing the starboard float. The aircraft began to vibrate, and therefore set course for base, after firing about 1,075 rounds. No results could be observed in the darkness. Position 3725N 0257E.

The *Luigi Torelli* managed to reach Bordeaux. The Squadron maintained its record a few days later:

> At 0834 on June 11, Sunderland W/10 [W3993, Flt Lt E. Martin] flying at 2,000ft, in cloud and rain (using the beam aerials) saw a U-boat 5 miles ahead, in a patch of better visibility; it was making 8kt. It was still on the surface when six depth charges were dropped up the track from 30–50ft (250lbs torpex, set to 25ft, actual spacing 40ft). They exploded all round it, and on their subsidence it was seen to be lying almost stationary in the centre of the disturbed area, with a list to starboard. It porpoised slowly (first bows up, then stern up), turned sluggishly to port and eventually gradually submerged, three or four minutes after the attack. A minute later it reappeared and opened fire with cannon. The Sunderland, returning fire, immediately attacked up the track from 600ft, dropping an anti-submarine bomb, but it fell short. Another was then released, and this exploded alongside. A great patch of oil then appeared; in spite of a rough sea which rapidly broke up the patch, it maintained a width of at least 50yd. The aircraft stayed 3½ hours keeping contact by Special Equipment in bad visibility, while the U-boat remained on the surface, moving slowly; its speed now varied between two and four knots, and occasionally it stopped, while its course was erratic, with a variation of 20 degrees. In all 700 rounds were fired from the aircraft, which was itself hit in the port wing. Position 4320N 1233W.

Date	*Aircraft*	*Captain*	*Time*	*Details*
Aug 3	W3989	W/O Ward	0556–1740	Escort 'secret force'
Aug 4	W4026	Flt Lt Goyen	0328–1542	Escort Cv SC93
	W3995	Flt Lt Bannister	0822–2013	Escort Cv SC93
	W4017	Flt Lt Maynard	1615–0543	Escort 'secret force'
	W4032	W/C Burnett	0448–1615	Escort 'secret force'
Aug 5	T9112	Flt Lt Briscoe	0400–1043	Escort Cv RV34
	W4032	W/C Burnett	0855–2105	Escort Cv RV34
Aug 11	T9112	Flt Lt Bannister	1025–1807	Escort Cv ONS120
Aug 12	W4032	Fg Off Fife	0530–1700	Escort Cv SL117
	T9112	Flt Lt Bannister	1403–2050	Escort Cv SL117
Aug 13	W6004	Flt Lt Maynard	1714–?	Escort Cv OS37
Aug 16	W4032	Fg Off Fife	0625–1722	A/S escort
	W4026	Flt Lt Goyen	1135–1737	A/S escort
	T9112	Flt Lt Bannister	0700–1355	A/S escort
Aug 18	T9086	Flt Lt Lovelace	0535–1630	Escort Cv SC95
	W4032	Sqn Ldr Flint	0705–2016	Escort Force LT
Aug 19	W4017	Fg Off Church	0858–0947	A/S escort
	T9112	Flt Lt Bannister	1129–2147	Escort Cv SN73
Aug 20	W4026	Flt Lt Goyen	1256–1925	Escort Force LS
	T9086	Flt Lt Lovelace	1047–2320	A/S search
Aug 27	T9086	Flt Lt Lovelace	1158–1215	Escort Force LX
Aug 29	T9086	Flt Lt Lovelace	0218–1415	Escort 'secret force'
	T9112	Flt Lt Briscoe	0823–2025	Escort 'secret force'
Aug 30	W4017	Fg Off Church	1910–0909	A/S search

The crew were credited with causing damage to the Type IXB U-105.

Despite having formed at Oban in May 1942, the Canadian 423 squadron did not fly its first Sunderland operation until 3 August, Fg Off J. Musgrave (W6053) undertaking an anti-submarine search.

August saw 10 Squadron continue its good run against the German U-boats, making eleven sightings and six attacks. Of the other Coastal Command Sunderland units, 461 Squadron made three sightings and one attack and 202 Squadron five sightings and three attacks. On 10 August 461 Squadron made this attack:

> At 1413, E/461 flying at 2,500ft sighted a U-boat on the surface 5 miles away on port quarter. The U-boat left a definite oil streak about 10yd wide and 1½ miles long. The aircraft turned towards the U-boat, which submerged when 2½ miles away, but when the aircraft was about 600yd from the swirl the conning tower and stern re-appeared for a short time. The aircraft attacked from the U-boats's port beam at right angles to its track and released six torpex depth charges (set to 25ft, spaced at 35ft) from 50ft, 12 seconds after the second disappearance of the conning tower. The DCs straddled the line of advance at the point of aim, about 60yd ahead of the swirl, which should have been the actual position of the conning tower. Five minutes after the attack an oil bubble 60 x 25yd and two large air bubbles were seen 200yd ahead of the explosion mark. This well executed attack certainly shook the crew severely, and probably caused a certain amount of damage. Position 4439N 1221W.

However, yet again, the Coastal Command summary of U-boat scores gives no official credit.

August was a typical month for 228 Squadron (*see* table left).

On 1 September the Austrialians were in action again:

> On the morning of September 1, R/10, U/10 and A/461 were engaged on an anti-shipping patrol in the Bay of Biscay. While proceeding to the patrol area, U/10 got a Special Equipment contact 12 miles on the port beam, and on homing sighted, at a range of 5 miles, a vessel that was emitting so much smoke that it was thought to be a merchantman. Approaching up sun and making use of cloud cover, the aircraft finally identified it as an Italian submarine, travelling at 6kt. The submarine opened fire with light flak from the after end of the bridge, but at 1028 hours the aircraft pressed home the attack from

A pristine-looking ML765 wearing prototype markings. Peter Green Collection

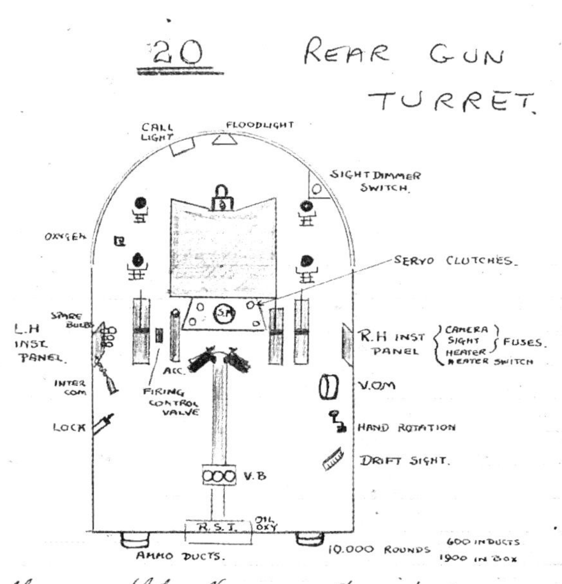

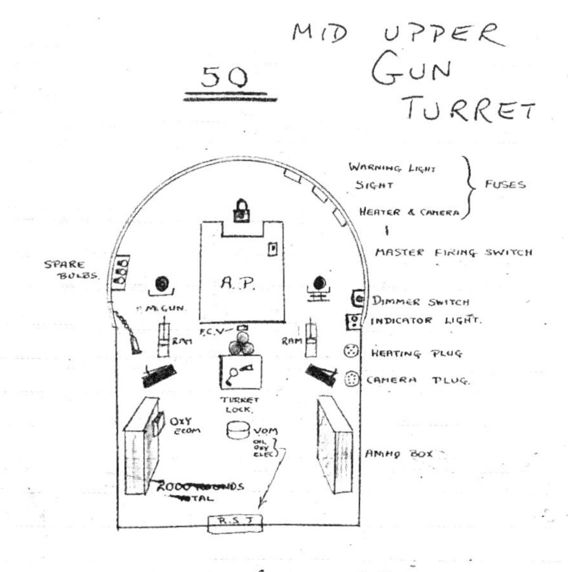

Rear gun and mid-upper gun turrets as sketched by Ron Gaunt in his notebook during the conversion course. Ron Gaunt

the port quarter, releasing a stick of four 250lb SAP bombs while the submarine was still fully surfaced. Only one bomb was seen to explode, and this about 30yd to port of the submarine, but yellow smoke immediately issued from its port quarter and continued for about half a minute. At this point, R/10 came on the scene, having obtained simultaneous Special Equipment and visual contacts at 10 miles' range, and carried out two machine-gun attacks on the submarine from stern to bow, diving from 1,500ft to 500ft and firing from nose and tail guns. The submarine replied from all gun positions, the cannon fire being intense. A few minutes later, R/10 sighted A/461 one mile away.

Meanwhile, U/10 had turned to port in a wide sweep and now attacked again from the starboard bow but, owing to an error, the bomb did not release. Eight minutes later R/10 attacked from the submarine's starboard beam with two 250lb SAP bombs, turning to repeat the same manoeuvre with one more bomb. None of the bombs fell sufficiently near the submarine to do appreciable damage. After R/10's second bomb attack the submarine ceased fire and did not fire again owing to casualties caused by the aircraft's tail guns. R/10 sustained several hits but suffered no casualties. Throughout the action a large volume of bluish-brown smoke came from the submarine's diesel exhausts, clouds of it trailed astern for half a mile.

The three aircraft on the scene then contacted each other by R/T and from 1035 to 1055 circled the position, arranging a concerted attack, but before this could be put into effect, orders were received from base to continue with the anti-shipping patrol. The submarine could consider itself lucky that the aircraft were carrying anti-shipping armament and had a major objective which precluded the use of every bomb. Position 4440N 0605W.

The Italian submarine *Reginaldo Guiliana* had been returning from operations off Brazil and was damaged in these attacks, although eventually reaching Santander.

Sunderland 'R' of 10 Squadron was soon in action again:

On September 9, R/10 was again flying in the Bay, at a height of 3,500ft, when a weak Special Equipment blip was received at 18 miles on the port bow. At 10 miles the contact grew strong. R/10 continued to home just above the thin layer of cloud at 2,000ft, and at 5 miles' range descended through it to sight a dark grey U-boat right ahead. The pilot climbed back into cloud, and again broke through at 3 miles' range, then dived at 195kt, turning slightly through the 3/10 cloud. Finally breaking cloud at 400ft the U-boat was seen 1½ miles away on the surface, travelling at 8kt. The aircraft turned slightly to starboard, observing men on the bridge and one man running along the upper deck, and circled astern at a range of one mile, when it climbed to 900ft, waiting for the U-boat to submerge. When the submarine was one mile on the starboard beam, the U-boat began to dive and the aircraft turned and attacked from its starboard bow, releasing six torpex depth charges with spoiler nose and tails (set to 25ft, spaced at 21ft) from 50ft, while the top of the conning tower and the stern were still above water. The stick straddled the U-boat's bow abaft the stern, three DCs on either side; they should have exploded just before the conning tower. The explosions were particularly heavy, and two minutes later large air bubbles effervesced for five minutes over an area 50ft across, 50yd from the inside edge of the explosion mark and continued for a quarter of an hour, when the aircraft left to adopt baiting tactics. Nothing new was observed when it returned 40 minutes later. This model attack undoubtedly inflicted severe damage to the U-boat. Position 4608N 0843W.

Plt Off Arthur Richmond of 230 Squadron was awarded the George Medal in September 1942:

Plt Off Richmond was a member of the crew of a Sunderland aircraft which crashed into the sea in September 1942. He observed another member of the crew (the WOP/AG) who was surrounded by a pool of burning petrol and was unable to swim owing to a broken leg. Although suffering from lacerations of the scalp, right arm, left leg and left knee, Plt Off Richmond dived under the flames, pulling his comrade to safety and eventually succeeded in getting him to a semi-inflated dinghy. He then supervised the rescue of three more of his companions and it was not until it became certain that no other survivors were in the vicinity that he abandoned the search and boarded a rescue boat. Plt Off Richmond displayed a very high standard of courage and devotion to duty and undoubtedly saved the WOP's life.

September 1942 saw the first Sunderland III, PP176, roll out of a new Shorts' facility on the shores of Lake Windermere; production facilities were in limited supply but the arrangement here was to be strictly temporary. Between this date and May 1944, the Windermere factory built thirty-five Sunderlands, although it also undertook an appreciable amount of overhaul and maintenance work on flying boats. For the U-boat war it was a similar picture in September, with 10 Squadron recording five attacks in six sightings, whilst 461 Squadron, now fully operational out of Hamworthy (Poole Harbour), attacked all three of the submarines they found.

The only confirmed success of the month, and indeed the first confirmed sinking by Sunderlands since July 1940, went to 202 Squadron on 14 September:

R/202 Squadron was off the coast of Algeria, flying at 800ft, when at 1430 hours an object was sighted 5 miles away. At 2 miles' distance this was identified as a submarine painted grey and green. The aircraft approached from astern so that the enemy's main gun could not be brought to bear, but there was some light flak, which ceased when the aircraft's front gun opened up. An attack was made from 50ft and five torpex depth charges were released while the submarine was still fully surfaced. One hung up, and of those which dropped, one took the fuzing link with it. Two DCs fell on the starboard side, just forward of the conning tower, and the other two along the port bow. Immediately the submarine lost all way, and oil gushed out all round it. It then steamed slowly round in circles, keeping the gun trained and firing at the aircraft, until 35 seconds after the attack, when it gradually sank bow first. About forty survivors were left floating in the water or on their dinghy. Position 3728N 0434E.

The Squadron ORB also contains an account of this attack:

U-boat sighted visually on surface, aircraft made circuit, coming in to attack on course of 230 degrees, aircraft took evasive action as U-boat opened fire with heavy flak when aircraft was two miles away. Aircraft attacked at 1434 with six Torpex depth charges, one of which hung up and another dropped with fuzing line attached, four explosions seen, two against port bow and two within 5ft of conning tower, starboard side. U-boat was still firing. Aircraft circled out of range and saw U-boat lift about a foot out of the water, and then circled to port, forward gun trained on aircraft but obtained no hits. This continued for about 30 minutes and the U-boat was slightly down by the bows. During this time some of the crew came up on deck. At 1507 U-boat started to submerge with crew still on deck. This was thought to be a *ruse de guerre* and aircraft made for U-boat with intention of dropping bombs. At 1510 some of the crew were seen taking to a dinghy and some jumped into the water. U-boat then sank bows first with stern sticking vertically out of water.

By September 1942 the average of U-boat kills from air attacks had fallen to 6 per cent per month despite improved tactics of search and final approach which, coupled with the gradually increasing number of aircraft devoted to anti-U-boat work, had resulted in a gratifying increase month by month in the number of attacks on U-boats. The insistence on better standards of ASV performance enabled unseen cloud approaches which, aided by white camouflage, enabled a large proportion of these attacks to be delivered on Class A targets – those in which the whole or part of the U-boat was still visible at the instant of depth charge release. Unfortunately, lethality of attacks had not risen anything like the figure hoped for after the introduction of Torpex-filled depth charges. (AHB *Narrative* 'The U-boat war')

Oban in late 1942 with DP198 of 422 Squadron. This aircraft was subsequently converted to Mk V and remained in use to 1959. Andy Thomas Collection

The prospect of equipping with Sunderlands was not always greeted with delight; when 119 Squadron was informed in September 1942 that it would receive a complement of five Sunderlands, the CO, Wg Cdr Gordon, went to Coastal Command HQ to request that the squadron keep its Catalinas: the answer was a firm 'No' and the Sunderlands duly arrived. It was, however, to prove a miserable few months and the unit was reduced to number status in April 1943, having achieved little with its aircraft and having suffered a number of problems; indeed, although the official documents are not very enlightening, there is a suspicion that the squadron was perhaps 'disbanded' for morale reasons. The unit had flown its first operation on 20 November, Fg Off McCallum in DV971 carrying out a convoy escort (this crew also became the Squadron's first combat loss, failing to return from a sortie on December 15).

A few days later, on 25 November, sister aircraft DV972 was on a depth charge test when:

> ... one exploded on hitting the water, damaging the starboard float, smashing the aileron and elevator, and putting numerous holes in the hull and starboard wing. The aircraft practically stalled on the water and the throttles were closed. The aircraft sank rapidly, in less than five minutes, and the crew took to two dinghies. (119 Squadron ORB)

An ASR Walrus landed to pick up survivors but was then too heavy to take-off again – and was also in a minefield area! An ASR launch subsequently collected the Sunderland crew.

There were always dangers lurking under the water. Flt Lt Lovelace took 228 Squadron's T9086 from Oban to Vaagar to deliver equipment to this remote Faroes site; the aircraft landed safely but then struck an uncharted rock in Sorvaag Vatn and had to be beached in 9ft of water with three holes in the keel. (The squadron lost another aircraft in December, W6004 being sunk at its moorings in Wig Bay after colliding with a Catalina which had broken from its moorings; the Sunderland had gone to Wig Bay from Castle Archdale for compass swinging. The bad luck continued into early 1943, 'P' sinking after running aground on the appropriately named Troublesome Rocks, Lough Erne.)

Meanwhile, another Sunderland squadron joined Coastal Command's ranks, 246 Squadron at Bowmore receiving its first two aircraft from Wig Bay on 11 October. The following day both aircraft deployed to Lough Erne on a fourteen-day training detachment, whilst towards the end of the month three crews collected more aircraft from Pembroke Dock. The squadron flew its first operational sortie on 12 December, Fg Off Squires (DV978) escorting convoy ONS152.

In the same month, 422 Squadron at Lough Erne (a Catalina unit formed in April 1942) was declared non-operational pending re-equipment with Sunderlands. The squadron moved to Oban in November, receiving its first two aircraft (W6028 and W6029) on the 9th, although both went straight to Wig Bay for 'operational fitting'. The first fully operational aircraft, W6026, was back at Oban on 17 November and the Squadron commenced training.

> A good beginning has been made on crew training, but due to the fact that very few of the aircrew personnel have previously had Sunderland training or experience, it is expected that the training period will of necessity be fairly lengthy. (422 Squadron ORB, December 1942)

The training period was marred by the loss, on December 19, of W6029.

In the latter part of 1942, as the U-boats continued to cause havoc with convoys, the Sunderlands were often called upon to mount ASR missions. Ken Robinson was the navigator in Fg Off Gall's 201 Squadron crew and he recalled one such mission:

> We left Castle Archdale for a creeping line-ahead search at 0450 and, at first light, descended through cloud to see a lifeboat with twelve occupants. Those in the lifeboat fired a red Verey and flashed SOS, then signalling that they were living on tea. This boat was one of four from an MV sunk by a submarine. The U-boat commander had interrogated them before leaving them adrift in mid-Atlantic. We assembled such food as we could from the galley and wrapped it in a Mae West. I manned the bomb-sight and we made a low-level 'bombing' run over the boat. The package was picked up by the occupants of the tiny craft who flashed their

thanks. The seas were too heavy to land and pick them up so the position was accurately recorded whilst Sunderlands, Catalinas and Flying Fortresses continued to search the area until four lifeboats were found. Until the naval destroyer came to pick them up, a constant watch was kept over the boats and food, water and medical supplies were dropped. In all seventy-three passengers and crew were saved. A great operation. (From *On the Step*, *ibid*)

The Battle of the Atlantic was to climax in 1943 when, at last, the Allies, largely due to the application of air power, reduced to 'acceptable and sustainable levels' shipping losses to German U-boats, whilst the U-boat force itself suffered crippling losses. Not all submarines engaged turned out to be German. On 5 January, a 204 Squadron aircraft, N9044 (Flt Lt H. V. Horner) was conducting a creeping line-ahead search on behalf of Force FG, which had suffered a number of U-boat attacks. The crew spotted a submarine and three escort vessels and when these failed to acknowledge an Aldis challenge, the Sunderland made a depth charge attack on the submarine. The nearest of three depth charges fell 30ft (9m) short of the starboard bow. The Sunderland re-positioned and closed again, at which point the submarine changed course

Heavy Seas

A Sunderland of 423 Squadron made a forced landing twelve miles west of St Kilda and took off again in a heavy sea. The Captain's account was included as an 'instruction to others' in an issue of the *Coastal Command Review*.

> We were recalled from convoy escort owing to adverse weather conditions. Prior to the engine failure I was operating the SE and on a signal from the second pilot I took over. Both the port engines had cut. At this stage we were about 500ft. I could not maintain height on the starboard engines. We sounded 'bombs out' and the emergency signal. At 200ft, being almost into wind, I throttled back and glided down, flaps were not used. Our landing was successful and there was no apparent damage to floats or hull.
>
> During these few seconds of landing we prepared to abandon the boat. Dinghy drill was completed, bombs rendered safe, and depth charges jettisoned on the water. We did not have time to do this in the air. The navigator completed his log and checked up the dinghy preparations. All the time we were on the water the radio operator was sending out an SOS and the position on four frequencies. Just after touching down we managed to restart the port engines. Although darkness was falling we could see St Kilda as we rose on the crest of the waves, and we tried to taxy towards it. The sea was too rough, and progress was slow, if any. Whenever we headed the boat out of the wind the starboard float submerged and the outer airscrew was chewing the sea.
>
> Heavy seas continually broke over us, water pouring in through the front turret, pilots' windows and astro hatch. One of our front windows was cracked. We were amazed at the punishment our Sunderland was taking. The boat was pitching and plunging violently – it is little wonder two or three of the crew were sick. After about half-an-hour of this we realized that we were shipping more water than we could pump out, and sooner or later the boat would sink. We could get no reply to our SOS, and not knowing whether it had been picked up we were reluctant to abandon and take to dinghies. Launching these would have been hazardous if not hopeless. We should not have survived long in that sea, which was becoming more violent. We were obviously getting no nearer to St Kilda, and the situation was not encouraging. At this stage I asked the crew how they felt about my taking off. We agreed it was worth it.
>
> The crew remained at dinghy stations. Our first attempt was along the length of the swell. This was unsuccessful, the wind was too strong. I tried opening the four throttles together, but a wave doused the inners. Finally, leaving the outboards wide open, and following up with the inners, we managed it. A bucking bronco had nothing on the Sunderland in that take-off. One second we were surfing down the backside of a swell burying our nose in the trough, the next we were bounced into the air. The pounding the boat took was phenomenal. The floor boards were buckling, crockery was flying around the galley with one of the navigators vainly trying to catch the cups as they ricocheted off the walls. On each bounce the boat gained a little more airspeed, finally becoming airborne about 65kt.
>
> So heavy was the beating out boat had taken, that the bows were bent upwards, the sides of the hull creased and longerons strained from the bows through to the galley bilge. The tailplane was bashed in on the leading edges and underneath, and the airscrews were pitted and bent. We had been on the water nearly an hour. Our thanksgiving feast consisted of hot tea and chocolate.

Two hours later the Sunderland landed safely at its base.

EK581 was a Blackburn-built Mark III and served with 423 Squadron at Castle Archdale. Peter Green Collection

and raised the French Tricolour; it was subsequently identified as the *La Sultane*; interestingly, the attack – which appeared accurate – caused no significant damage.

Other than the standard markings and two-letter squadron codes worn by UK-based squadrons, Sunderlands carried no other national insignia; however, in late January 423 Squadron added a 'Canadian touch' to their aircraft, painting an 8in (20cm) roundel in RAF colours on the port side just in front of the wing – but adding a maple leaf to the centre circle.

The winter of 1943 brought its own hazards for the squadrons stationed in Scotland and a number of Sunderlands were damaged or wrecked in winter storms; the 246 Squadron ORB records the problems at Bowmore. A gale warning was issued on 5 February and at 2230 the engines of the aircraft were started as winds increased to 50mph (80km/h), rapidly rising to 90mph (150km/h). Sunderland 'L' was reported adrift and although it attempted to weather the gale it was in danger of floundering, the port float having filled with water, and so the crew beached it. A week later the situation was even worse, another gale warning was issued and crews started engines at 0730 on 12 February:

> W6057/B had main pennant broken, W6058/C was ordered to jettison petrol in an attempt to raise the port float which was now badly down. A very bad sea was running, W6060/D reported that strap from buoy had broken. 'B' and 'D' now riding out the gale on engines only. At 1430 'C' condition serious and therefore slipped cable and beached at 1545 on soft sand. At dawn on February 13 the gale abated enough to put fresh gale guards on board. 'C''s crew had been aboard forty-five hours. By midday on the 14th the gale had picked up again and engines were kept running. It finally abated on the 16th and crews were brought ashore. (246 Squadron ORB, February 1943)

All three aircraft survived their ordeal, as did the crews. Appendix I gives details of each Sunderland and the fate column includes numerous references to 'sank in gale', No 4 OTU at Alness seeming particularly vulnerable.

The Norwegian 330 Squadron had moved from Reykjavik to Oban in January and the following month its Northrop N3P seaplanes were joined by the first Sunderlands. After six weeks of conversion flying, and the arrival of more Sunderland IIIs, the Squadron flew its first Sunderland operation on 20 April, Lt P. Thorendahl (W6007) providing cover for convoy ON179. March had brought the operational debut of another Canadian Sunderland unit, 422 Squadron commencing operations from Castle Archdale (the squadron had actually flown a number of sorties in January whilst on attachment to No. 19 Group).

A Coastal Command memo of February 1943 stated that another Sunderland squadron was to be formed 'in the near future', a pencil addition to the memo adding '190 Squadron', although this squadron actually formed with Catalinas. Later in the month, on 26 February, the C-in-C Coastal Command wrote to ACAS(P):

> Under my proposals [for Force H] three flying boat squadrons – 422, 119, 246 squadrons – will become surplus to this Command and will nominally be available for overseas if required. In point of fact the majority of personnel and aircraft will have to be absorbed by the remaining squadrons to bring them to the full new establishment.

This latter comment reflected the planned change to twelve aircraft per squadron. Amongst the Coastal Command documents referring to the various force projections, an appendix to the Target Force H document is particularly interesting in that it includes the following performance table. It listed all Coastal Command types but I have extracted the three Sunderland references, plus two other appropriate aircraft – Catalina I and Liberator III – for comparison (*see* table below).

One of the major reasons for the difference in patrol time was the better performance of the American engines over the thirsty Pegasus engines of the Sunderland, a point made as early as 1940 during the discussion on Sunderland replacement. Indeed, it would have been interesting to add the Sunderland V (Pratt &Whitney engines) performance figures to the table below.

Following the March 1943 Atlantic Convoys Conference in Washington, revised strategies were implemented in terms of convoy organization, air patrols, and making a concerted effort to cripple the U-boat force. To effect this latter plan, operations were intensified in the Atlantic and, more especially, in the Bay of Biscay.

The air offensive against the U-boats intensified with particular attention being paid to the Bay of Biscay in order to catch the submarines on the surface as they transited to and from their bases in France. The new patrol areas included 'Musketry' and 'Seaslug'. Coastal Command had issued an amendment on 14 March to its tactical instructions, increasing depth charge stick spacing to 100ft (30m), various trials having concluded that this was the most effective way of combing coverage with lethality. This change is not, however, reflected in the attack reports submitted by Sunderlands, most of which still refer to 60–75ft (18–23m) stick spacing. Late March saw Coastal Command activate Operation *Enclose*, a concerted effort against U-boats in a specified north-south ribbon over the Bay of Biscay. Three Sunderland squadrons (10, 119 and 461) took part in this operation, between 20–28 March as part of a 115-aircraft force. The air patrols flew 1,300 hours during this period and made twenty-six sightings, fifteen resulting in attacks, Leigh Light Wellingtons managing to sink U-665 and damage U-332. The sightings record was excellent, considering that only forty-one U-boats crossed this area during this period.

On 19 March Flt Lt G. Church of 228 Squadron (DD837/V) attacked a U-boat, a second one being attacked by 201 Squadron's Fg Off W. Robertson the following day; the latter attack has been a cause of some confusion, credit being given for the destruction of U-384 in official records or damage to U-631 in other accounts. Both these missions were connected with convoy patrols for HX229 and SC122. According to the RAF's post-war analysis of U-boat fates, U-384 was sunk by a 206 Squadron Fortress on 20 March.

	Speed	Load	Op Range	Endurance	Hours on escort at range from base (sea miles)						
	(mph)	(lb)	(sea miles)	(hr)	300	400	500	600	700	800	900
Sund. II	110	2,000	1,270	11.55	6.1	4.3	2.5				
Sund. III	110	2,000	1,313	11.95	6.5	4.7	2.8				
*Sund. IV**	132	2,000	1,960	14.8	10.3	8.8	7.3	5.8	4.2	2.7	
Cat. I	100	2,000	1,840	18.4	12.4	10.4	8.4	6.4	4.4	2.4	
Lib. III	145	1,500	2,300	15.9	11.3	10.0	8.7	7.3	6.0	4.7	3.3

*Estimated figures

July 1943 and an aircraft of 330 Squadron is taken in tow by a RAF High Speed Launch (HSL). The Norwegians received their first Sunderlands in February 1943.
Ken Delve Collection

Two aircraft of 119 squadron were airborne on 29 March flying ASPs; on return, DV179 was diverted to Mount Batten but the balloon barrage 'hindered the correct landing approach and the aircraft overran, hitting rocks. The hull was holed and a float torn off. The aircraft being submerged to mainplane level.' Two days later, Flt Lt Wharram had the port inner prop fall off W6002. The same day Flt Lt Gibson (PP176) was tasked to intercept and shadow a blockade runner; having found the vessel he elected to make an attack but was forced to break off in the face of heavy flak.

Coastal Command decided to run similar barrage operations during April. Operation *Enclose II* covered much of the same area from 5–13 April, although only eighty-six aircraft were involved, including ASV II-equipped Sunderlands, much of the effort being night sorties. Eleven U-boats were detected but only four attacked – although one was sunk (U-376 by a 172 Squadron Wellington) and one damaged (U-465). The Sunderlands detected five submarines but, lacking suitable illumination, were only able to make one attack. For the remainder of the month the patrol area was extended and more aircraft involved; as Operation *Derange* this continued into late May. During the April period, the RAF flew over 2,500 hours on *Derange* and detected thirty-six U-boats, twenty-two of which were attacked. Only one (U-332) was sunk, in early May, although two were damaged to such an extent that they had to return to France for repairs. The campaign was proving effective and on 1 May Admiral Dönitz issued his 'Fight Back' order (Standing War Order 483) to the U-boats, whereby the submarine should engage the attacking aircraft rather than risk an emergency dive which would leave it vulnerable to depth charge attack. The stage was set for a battle that Coastal Command were happy to accept.

Despite this concentration of effort over the Bay of Biscay, operations over the Atlantic also remained a standard part of the overall tasking. The Sunderland's endurance on these AS patrols was around twelve hours, and at the Sunderland operating speeds this was not enough to cover the so-called Atlantic Gap: only the limited number of Very Long Range (VLR) Liberators were able to cover this dangerous area. Various ideas were put forward to try and increase the Sunderland's operational range; Flt Lt Hewitt of 201 Squadron was instrumental in that unit's attempts to improve the situation, the solution they adopted being one of removing as much 'unnecessary' weight as possible from the aircraft. With a fuel load of 2,500gal (11,000ltr) this had the effect of increasing the aircraft's endurance to fifteen hours (plus a three-hour safety margin); however, this was modified a little while later to a fourteen-hour operational endurance using a 2,250gal (10,000ltr) fuel load. The 2,500gal limit was re-introduced in May and one aircraft recorded a sortie of 18 hours 40 minutes – a record for the Sunderland.

As briefly mentioned above, April had been no better for 119 Squadron. Having departed Pembroke Dock at 1955 hours on 14 April, Flt Lt J. V. Gibson and 11 crew Sunderland DP176/D were tasked with a Flooder Patrol over the bay. At 0024 hours the captain elected to return to base as the port outer exhaust ring was holed.

> At 0210 the port outer and inner airscrews flew off without warning at height of 4,000ft. W/T operator continued sending SOS. 0211 hours aircraft ditched, losing port float and port wing-tip. Aircraft buried and stood on its nose and remained in this position for about ten minutes with water level with mid-upper turret. All members of crew left aircraft within twenty seconds of ditching. Only two dinghies were got out and tied together. They were separated and one was floated towards three members of the crew who were clinging to aerial, but it floated away unopened with current and they did not get into it. Two of these three reached other dinghy; the remaining one, Fg Off Waters (Navigator) who was being assisted over to the dinghy, broke away and disappeared below the water and was not seen again. The remainder of the crew got into dinghy, except Flt Lt Davies (tail gunner) and Sgt Galloway (Flight Engineer) who were not seen after they left the water. The only equipment salved with the dinghy was two marine distress signals and one Verey Pistol containing a cartridge (the latter wet and u/s). At 0330 all nine remaining crew on dinghy – all alright except Sgt Parker (2nd Pilot) who was delirious. At 0805 Sunderland 'O' arrived and remained with the dinghy. About 1200 Catalina 'L' arrived and both a/c remained in company. Sunderland dropped within 25ft of dinghy, as dinghy drifted faster than the bags. At 1235 Sunderland sent 'V/S ship approaching now' and left vicinity. At 1510 hours, D/R *Wensleydale* arrived and picked up all survivors. (119 Squadron ORB, April 1943)

On April 15 the squadron's aircraft were transferred to other units.

By spring 1943, No. 4(C) OTU was turning out an average of twenty Sunderland crews a month from each ten-week course. The strength of the unit had stabilized at seventeen Sunderlands (plus nineteen Catalinas). However, the move from Invergordon to Alness was not popular with the unit's Commanding Officer, Gp Capt Gordon, and on 12 April he wrote to AOC No. 17 Group:

> The geographical situation of the Station and the meteorological conditions prevailing in this area are such that the expansion of this Station from a temporary base to a large Flying Boat OTU from the point of view of operations was a most unfortunate decision. The subsequent building of a widely dispersed Station is regrettable. Dalmare Pier is the operational pier for the Servicing Squadron and Flying Flights; this pier is so situated that it cannot be used except under reasonably calm conditions i.e. not more than 20mph. In rough and heavy weather the Servicing Squadron and flying crews have to be transported by bus from the Technical site to Invergordon and ferried to and from aircraft from there.
>
> The position of the moorings, consisting of two trots on each side of the Firth, each 2 miles long, situated in a narrow area completely exposed to the prevailing wind could hardly be worse. A boat guard [gale guard] of two aircrew is placed on each flying boat before dusk with rations and full instructions on the inspection of the hull and prevention of leaks. On receipt of a gale warning it is thus only necessary to put on one pilot per boat.

The boat guard requirement was removed a little while later as the use of aircrew in this way was having an impact on the training course. Alness certainly suffered a high aircraft accident rate – but then so did most training units. Meanwhile, from late April, the other flying boat training unit, No. 131 OTU, had been increased to eleven crews per course, Sunderlands being added to the inventory.

Slessor wrote to one of the disbanding units, 246 Squadron (Wg Cdr Laws) on 24 April:

> I would like you to know that I share the disappointment which I'm sure you are all feeling at the disbandment of 246 Squadron. We all know that the Sunderland has its good points, but in this stage of the war when we have to put out a greater effort and also to economise in manpower there is no alternative to cutting down overheads in organization wherever possible. So we have had to reduce the number of squadrons and correspondingly to increase the establishment of those remaining.

The Squadron disbanded on 30 April (it eventually reformed in October at Lyneham as a Liberator unit). The third of the squadrons mentioned in the February letter, 422 Squadron, remained in existence as a Sunderland unit, there being various agreements with the Canadian Government that would have made the disbanding of this unit politically tricky.

Sunderland crews were always ready to take risks in landing on the open sea to rescue survivors in dinghies, the standing Coastal Command policy being 'land at discretion'. The history of the aircraft is sprinkled with such rescues and it is important to realize the part such actions played in the history of the aircraft. There are also frequent instances of when it all went wrong; such was the case on 28 May for JM675 (Flt Lt Dobs) of 461 Squadron:

> The aircraft flew over the dinghy containing six survivors of a 10 Squadron Whitley, it attempted to land but bounced three times on the

Two views of a dramatic landing on the airfield at Angle; the flying boat had been badly holed in the forward part of the hull and the pilot elected to make a landing on the grass airfield. Ken Delve Collection

uneven swell, diving vertically into the fourth swell. The pilot was killed but the rest of the crew escaped into a dinghy.

The following day, Plt Off Singleton (T9114) from the same squadron landed near the dinghies; the survivors were subsequently transferred to the French destroyer *La Combattante*; 'when trying to take-off with a skeleton crew the aircraft was struck by a huge swell that put a 7 × 4ft (2.1 × 1.2m) hole in the hull. Realizing it would be suicidal to attempt a sea landing the captain signalled that he was landing at Angle aerodrome. A perfect landing was made with no further damage.' (461 Squadron ORB, May 1943).

Battle over Biscay Intensifies

April 29 brought the claimed destruction of a U-boat, attacks being made by 'F' of 10 Squadron RAAF and 'P' of 461 Squadron. The 10 Squadron aircraft was first to attack:

At 1102 hours a smoke float was sighted; the aircraft investigated and then sighted the periscope of a U-boat in the act of surfacing. S/E was switched off. The U-boat opened fire and the aircraft attacked from the port quarter, releasing six Mk VIII Torpex depth charges, Mk XVI pistol, set to shallow depth, spaced slightly less than 100ft, while the U-boat was still on the surface. Immediately after the depth charge explosions, an oil streak was seen and then blue and black smoke.

At this point the 461 Squadron aircraft appeared and carried out an attack, its depth charges being accurately placed, although no success was credited. Two days later this same captain, Flt Lt E. Smith, was airborne in DV968 and in an attack on U-415 damaged the U-boat. The following day he and his crew, in the same aircraft, were even more successful, sinking U-332:

1038, SE contact. Sighted the conning tower of a U-boat and dropped a flame float, aluminium sea marker and marine marker. 1124 sighted U-boat on the surface and attacked with six depth charges. The U-boat sank horizontally and then the stern emerged and disappeared vertically, one or more sailors were seen in the water. (461 Squadron ORB)

Attempts by U-boat planners to re-route away from areas of intense air patrols met with only limited success. The U-boats often transited in twos and threes, their basic tactic being to use concentrated gunfire to chase away any threatening aircraft.

The present U-boat tactic of staying on the surface and fighting back has created a new problem for the crews of Flying Boats. Statistics show that Flying Boats are by no means inferior as weapons for engaging aggressive U-boats, but, in spite of his unstable and narrow gun platform, the Hun does occasionally get in a lucky shot. Unfortunately, he always has an almost uninterrupted view of the Flying Boat's most vulnerable area, its bottom, or as the more technical people would say, the area below the chine.

This area has to sink into the water if a normal landing is to carried out and a certain amount of embarrassment is caused if the bottom is full of holes. After all, a colander will not float and neither will a Flying Boat when it is riddled with holes. However, we have several advantages over a colander as leak stoppers are provided to block up holes and an auxiliary power unit is at hand to pump water out as fast as it comes in, provided, of course, the holes are not too big. You may ask 'but suppose the holes are too big'. Then there is only one answer. We must land the boat on the water and run on to the beach or, better still, land on land, preferably a grass aerodrome.

It is obvious that no definite procedure can be laid down, as each and every case will be different. First, having, we hoped, demolished the U-boat, an inspection must be made. To do this all the floor boards must be taken up and a careful inspection made to see if any daylight is showing.

W3985 of 10 Squadron dropping a stick of depth charges on a U-boat, 19 April 1943 – no damage was credited. Ken Delve Collection

Having found the holes, jagged metal should be removed, or, better still, hammered back into place as this will reduce the size of the hole. Leak stoppers should then be inserted where possible. Remember that a hole in the bow has to take a weight of water at nearly 90 degrees, whereas a hole behind the main step takes a weight at a much less angle. Concentrate on the holes in front of the main step first, and, if after using all your leak stoppers there are still some holes to plug up, block these with something solid, wrapped round with rag or anything you can find. Keep one of the crew to stand by each particular hole if possible.

Have the APU ready and primed for when you land and two of the crew ready to go up top to start the motor for pumping. Do not forget to inform base of your intentions as they will then have a Marine Craft ready for you. Make sure that everyone has his Mae West on.

In one recent incident a Sunderland had been shot up by a Fw 200 and badly holed underneath. The holes in the hull were patched up as best as possible, some with leak stoppers and the rest with rags. The APU was unserviceable, but on landing, the Captain organized the Bilge Hand Pump and a Zwicki Pump which was available, and the water was kept at bay. Approximately half an hour after landing on the water there were many volunteers upon the jetty standing by to take a hand at the pumps. It was impossible to beach the aircraft that night as no beaching chassis was available, but the pumps were kept going until morning when the aircraft was brought up the slip. (*Coastal Command Review* II/2, June 1943)

The same document stated that landing on grass was not a problem:

A Sunderland recently pulled off a very neat landing on the grass part of an aerodrome and nothing happened. The aircraft was slightly bent, but all the crew walked out smiling. The only headaches were among the Maintenance Section. As maintenance people have your interests at heart they do not mind an odd headache, particularly if you use a little foresight from which it will be an easy matter to return your aircraft to the water.

Another U-boat (U-465) was sunk on 7 May, falling to Flt Lt G. Rossiter and his 10 Squadron RAAF crew. A few days later, on 12 May, the Canadians scored their first success, 423 Squadron's Flt Lt J. Musgrave (W6006) putting paid to U-456, this U-boat having been damaged in a previous attack by an 86 Squadron Liberator.

No. 228 Squadron had been declared operationally non-effective on 29 April as part of the planned move to Pembroke Dock; this, however, was short-lived and the squadron was soon back in action, the new hunting ground being the Bay of Biscay. May proved a mixed month, EJ139 (Fg Off H. Debden) failing to return from a *Derange* patrol over the Bay on 24 May, and a few days later, on 31 May, Fg Off W. French (DD838) gaining some revenge by sinking U-563. This was a shared victory with DV969 of 10 Squadron RAAF (Flt Lt Mainprize) and Halifaxes of 58 Squadron: Fg Off French's attack had delivered the *coup de grâce*.

The fact that this was a difficult and dangerous time for the U-boats is evident from the number of occasions on which the submarines were attacked by more than one aircraft. On 31 May two Halifaxes from 58 Squadron attacked with no definite result, although the target did appear to be down at the bows; at that point they spotted a Sunderland and one of the Halifaxes flew in its direction, flashing to attract its attention. The Sunderland, DV969/E of 10 Squadron RAAF (Flt Lt Mainprize), made its attack and

... [the] U-boat, which had been trailing oil and manoeuvring freely, stopped. The aircraft circled and made its second attack with four depth charges, two minutes later, from the starboard beam. After the second attack the U-boat was down by the bows, stern clear of the water. It appeared to be sinking slowly.

Whilst circling the scene, DD838X of 228 Squadron (Fg Off W. French) appeared, having been alerted by base to attack a damaged submarine. This Sunderland also made two attacks, both of which straddled the now stricken submarine. The U-563 had been operational since mid-1941 and was on her eighth war cruise, out of Brest, when sunk. The same day, U-440 fell to Flt Lt D. Gall of 201 Squadron.

Other moves at this time included 422 Squadron leaving Oban for Bowmore in early May, their initial reaction being that 'accommodation from the technical and personnel standpoints is inadequate at the new base'. The first month at Bowmore was also marred by the loss of DD846 (Fg Off E. Page), which crashed on 25 May near Clare Island with the loss of all crew.

May 1943 also saw the more effective 10cm ASV enter service. Work had been underway on the system since 1940 but the scanning system used in the prototype Wellington installation was unsuitable for installation in the Sunderland. Experiments with wing blisters progressed well and in December 1941 a development contract was placed with Metropolitan-Vickers for a double mirror synchronized scanning system. During February 1942 a Whitley had been fitted with an experimental model of the double mirror system and the 10cm ASV as a development stage before fitting to a Sunderland. Shorts considered that the arrangement was practical for the Sunderland; however, there was still no sense of urgency with the programme. A Coastal Command memo of 28 May 1943 stated that 'production of 10cm ASV was to be undertaken on the highest priority for installation without delay, both retrospectively and in aircraft production lines, in Sunderland, Wellington VIII and Whitley VII aircraft' ('Signals' *ibid*). Although trial installations were successfully carried out

Destruction of U-106 by N/228 and E/461 on 2 August 1943.
Ken Delve Collection

in Sunderland and Wellington VIII (Leigh Light), by mid-1943, the Sunderland installation had dropped to fifth priority for ASV Mk III behind Wellington and Halifax variants.

U-boat losses (to all causes) peaked in May, forty-one being lost (five to Sunderlands) – it was a decisive month. On 24 May 1943 Dönitz suspended attacks on convoys in the North Atlantic; there was some debate in German naval circles about making this a permanent ban, but with new weapons and tactics under development, and a war situation that meant the U-boat campaign had to continue, it was not long before the submarines were back in action in this region. One of the great hopes the Germans had was that increased firepower would enable the U-boats to fight off air attack.

New weapons did enter service; U-758 was the first to become operational with a new quad gun mounting but it was soon evident that it was not working.

> Enemy pilots had overcome their initial fear of the new anti-aircraft armament and were again relentlessly pressing home their attacks in the face of it ... the 2cm shell often appeared to be ineffective. There were many occasions on which as many as a hundred hits had failed to disable an aircraft, suggesting that enemy heavy bombers and flying boats were now more heavily armoured beneath the fuselage. (German Naval History)

However, the Germans also realized the importance and vulnerability of the Biscay routes and employed aircraft such as the Ju 88 in defence of the U-boats: this led to a number of combats in which the Sunderlands could be out-gunned and out-manoeuvred. Flak boats had been introduced to lure and destroy anti-submarine aircraft, the first success for these flak U-boats coming on 24 May when U-441 shot down 228 Squadron's EJ139 (Fg Off Debden), although the Sunderland also managed to damage the U-boat (*see* above). The U-594 also claimed to have shot down a Sunderland on this day. However, the flying boats were able to look after themselves when attacked by enemy fighters; in one such instance over the Bay of Biscay on 2 June, 'N' of 461 Squadron (Flt Lt C. Walker) was engaged by no less than eight Ju 88s:

U-boat taking evasive action whilst under attack from a 461 Squadron aircraft on 24 April. 461 Squadron

> The Sunderland was carrying out a normal A/S patrol when eight Ju 88s were spotted six miles distant on the port quarter. The Ju 88s were flying at 3,000ft in three formations, one of four and two of two behind. The Sunderland opened full throttle and made for what cloud cover there was (3/10ths at 3,000ft). The Ju 88s gave chase and took up attacking positions three on each beam, 1,500yd distant, and 1,500ft above and one on each quarter at the same height and distance. The Sunderland jettisoned its depth charges and prepared to meet the attack.
>
> The Ju 88s peeled off to attack in pairs, one from each bow. The first attack hit the port outer engine, setting it on fire, and also resulted in an incendiary bullet entering the P4 compass, setting alight to the alcohol. The engine fire was extinguished by means of the Graviner switch, but the engine became unserviceable. The alcohol fire, which had set the captain's clothing alight, was also put out with the fire extinguisher. In the meantime, the first pilot took control and continued with the evasive action. During the attack, one of the Ju 88s attacking from starboard broke away, exposing his belly to the midships gunner at point blank range. The midships gunner fired and the Ju 88 burst into flames and crashed into the sea, disintegrating immediately.
>
> The next attacks severed the hydraulics of the tail turret, shot away the elevator and rudder trimming wires and scored numerous hits on the hull. However, the midships and nose gunners made the enemy pay for his success and another Ju 88 went blazing down towards the sea. It tried to pull up and ditch but hit the water, bounced vertically and crashed nose first into the sea. Simultaneously another Ju 88 came in on the starboard quarter and his burst fatally wounded the starboard galley gunner. Shortly afterwards a Ju 88 came in on the port quarter and the midships and tail gunners opened fire, the latter, owing to hydraulic failure, depressing the sears with his fingers in short bursts. The Ju 88 crashed into the sea – leaving five Ju 88s to attack the badly damaged Sunderland.
>
> 'Conditions' in the words of the Form Orange 'now became chaotic'. The intercomm and the radio were shot away, the ASI ceased to work, the navigator was wounded in the leg by shrapnel, and evasive action was controlled by hand signals from the navigator to the second pilot and thence to the captain. Owing to the unserviceable engine and the damaged controls it required both pilots to carry out evasive action.
>
> A Ju 88 came in from the starboard bow to meet the fire of the nose gunner – it broke away with the port engine ablaze and with smoke pouring from the cockpit. The combat continued for forty-five minutes in all and the crew of

the Sunderland estimated that every Ju 88 had been hit, with three shot down. The Sunderland managed to reach England and beach itself with one member of the crew killed and another injured. Brilliant shooting, skilful evasive action and determination in the face of heavy odds had altered what looked like an easy kill for the German force into a disastrous hammering. (*Coastal Command Review* II/2, June 1943)

For this action Flt Lt Walker received a DSO, Fg Off K. Simpson (navigator) a DFC, whilst DFMs went to Flt Sgt Goode (rear gunner) and Flt Sgt Fuller (midship gunner).

Submarine U-564 was damaged on 13 June by Fg Off L. Lee of 228 Squadron, although the aircraft was shot down by flak (the U-boat limped back towards base but was attacked and sunk by a 10 OTU Whitley, although once again it shot down its attacker). Another 228 Squadron Sunderland attacked one of three U-boats it found the following day. Fg Off White (JM678) dropped four depth charges in an indeterminate attack; the aircraft was subsequently engaged by Ju 88s but escaped into cloud.

Promising targets could not always be dealt with; on June 17 Flt Lt S. Butler (W6031, 422 Squadron) was on a Sea Slug 3 patrol when he sighted three U-boats on the surface. While attempting to close, his aircraft was engaged by the main guns and 20mm fire from all three enemy vessels, and the best he could do was circle the area whilst his rear gunner sprayed the enemy (to little effect) and the wireless operator attempted to fix the transmitter so that other aircraft and ships could be summoned to this tempting target. It proved to no avail and the three submarines vanished into the haze.

Sunderlands scored one more success in June, 201 Squadron's Fg Off B. Layne (W6005) operating out of Castle Archdale when he destroyed U-518 on 27 June.

12 July brought another loss during a Biscay patrol, 228 Squadron's DV977 (Sgt R. Codd) failing to return. The following day, JM708/N (Fg Off R. Hanbury) of 228 Squadron was on a Musketry patrol when it spotted three surfaced U-boats nine miles away. The Sunderland was soon joined by a Halifax and both aircraft circled the U-boats, which made no attempt to dive.

31 May 1943, the destruction of U-563; EJ139 (Fg Off French) of 228 Squadron watches the attack by Flt Lt Mainprize of 10 Squadron RAAF. Ken Delve Collection

After some time the Sunderland managed to separate one U-boat from the formation and attacked from fine on the port quarter. Seven Mk XI Torpex depth charges, set to shallow depth, spaced at 60ft, were released from 50ft. Evidence states that three charges fell close to the port side aft, one on the conning tower and three close to the starboard side forward. The attack was made under heavy flak and the pilot had to jink over the conning tower after release. The tail gunner and at least one other member of the crew saw the conning tower blown into

Inter-Squadron Navigation Competition

In a blaze of initiative two squadrons in No. 19 Group recently challenged each other, through the medium of the Group Navigation Officer, to a show-down of navigation skill; or rather, 502 Squadron, a landplane squadron, considered that it was high time the traditional superiority of navigation in flying boats, such as those owned by 10 Squadron RAAF, was called into doubt. Six navigators of the former and four of the latter were selected and each submitted four consecutive logs for assessment by Group. As the Group Navigation Officer finds it impossible to award the laurels in either direction, we have been called in to mediate.

	502 Sqn	*10 Sqn*
Flying time	188hrs	150hrs
Mileage flown	22,565	17,158
Calculation error	1.61%	0.98%
Final error	1.27%	1.31%
W/V frequency – out	60min	60min
– back	105min	90min
Frequency of position lines	80min	60min

There were also 'marks' for log keeping – neatness, completeness, preparations.

The figures for final error equate to between thirty and forty miles in a ten-hour sortie, with something like a 15–20-mile error half way around. No navigator will agree that this is a good thing, but he will often be content with his navigation if he returns to his base without serious trouble, oblivious of the possibility of his not having covered the required area. However, this 'worse case' scenario of final error assumes all errors are cumulative and that they will add up vectorially in a straight line – this is not a normal situation. We consider that on the whole there is very little to choose between the two squadrons' navigational skill and lest anyone should be tempted to rest on his laurels, we propose to distribute the now withering leaves between all contestants, at the same time enjoining them to make even greater efforts in the future. (*The Coastal Command Navigation Review* No. 5)

Destruction of U-106 by N/228 and E/461 on 2 August 1943. Ken Delve Collection

the air. A large part of the bow moved forward, stood on end, went over the vertical and slid under the surface. (*Coastal Command Review* II/3, July 1943)

The aircraft was subsequently shadowed by a Ju 88 but was not attacked, the Sunderland becoming waterborne at 1635. Submarine U-607, a Type VIIC, had left St Nazaire on its fifth war cruise on 10 July, its mission being to lay mines off the coast of Jamaica.

On 15 July, 330 Squadron moved to Sullom Voe. This unit had suffered its first operational Sunderland loss on 5 June, Plt Off Lester failing to return.

Towards the end of July a number of U-boats set sail across the Bay of Biscay to their operating areas; Sunderlands were to account for five of these submarines between 30 July and 4 August. The first success went to Flt Lt D. Marrows of 461 Squadron (W6077), this aircraft, coincidentally, sinking U-461:

> The crew saw three U-boats on the surface and a Halifax and Liberator overhead. The Sunderland attacked the U-boat that the Halifax appeared to have hit, but was driven off by flak from all three boats. The Liberator had also been driven off but the Sunderland then attacked at the same time and straddled one U-boat with seven depth charges, machine-gunning the vessel at the same time. The U-boat submerged, froth and scum appearing; later, 20–30 survivors were seen in the water and a dinghy was dropped to them. The Sunderland then attempted to attack the lead U-boat but by then an arriving Escort Group had commenced shelling the target and so the aircraft broke away. The aircraft had been hit several times. (462 Squadron ORB)

The Musketry patrols continued to bring action for the Sunderlands. On 1 August Flt Lt K. Fry (W4020) of 10 Squadron RAAF observed five sloops and a Catalina engaged in a U-boat hunt and decided to join in.

> The enemy vessel was sighted about six miles from the sloop. She carried one big gun forward, one 20mm gun on the bridge and two 20mm guns on the single bandstand. The Sunderland flew over the U-boat and made a tight turn to attack from the starboard quarter at 60 degrees to track. During the approach the front gunner opened fire, but the aircraft was subjected to very accurate return fire from the 20mm gun on the bridge. First the aircraft's inner engine was hit and then, when the aircraft was about 400yd away, a hit in the starboard main fuel tank caused petrol to pour out on to the bridge. It is believed that all three pilots were seriously wounded. The attack was nevertheless gallantly pressed home, and six depth charges set to shallow depth and spaced at 60ft were released from 50ft. Three depth charges fell on either side of the target, and the rear gunner saw the U-boat lift out of the water and then sink by the

No. 461 Squadron formed at Mount Batten in April 1942 and remained operational with Sunderlands for the rest of the war. Ken Delve Collection

bows. After the attack the Sunderland maintained course for about six miles, turned 180 degrees to port and ditched down wind at about 45 degrees to the swell. Apparently the captain was trying to get as near as possible to the sloops. The aircraft bounced twice and then settled, with the hull very seriously damaged. Six members of the crew succeeded in getting out on to the starboard mainplane which had broken away from the rest of the aircraft. They used this as a raft for about half an hour and were then picked up by ship. Fourteen survivors from the U-boat, which sank in thirty seconds, were also picked up [by HMS *Kite*]. (*Coastal Command Review* II/4, August 1943)

U-454 was destroyed in this attack.

A 228 Squadron aircraft was also in action the same day. JM678 (Fg Off S. White) had departed at 1450 on a Musketry patrol and whilst tasked to search for a dinghy came across a U-boat:

At 2013 hours the aircraft lost height and began an attack from up sun. The U-boat put up very rapid fire and the aircraft took violent evasive action and thus failed to track over the target. During the approach the gunners opened up accurate return fire and two of the German crew were seen to fall. The Sunderland then turned to port and ceased fire. When the aircraft was 600yd away the enemy again opened very accurate fire which carried away the starboard float and put the starboard aileron out of action. The hull was holed in several places and a shell exploded in the port mainplane. There were, however, no casualties in the aircraft. The attack was delivered from the U-boat's starboard quarter at 15 degrees to track; seven depth charges, set to shallow depth, spaced at 60ft, were dropped from 75ft. The depth charges straddled the target just abaft the conning tower and the U-boat was completely enveloped in spray. When the plumes subsided the U-boat had a bad list to port and was turning sharply. The Sunderland left the scene immediately, as the damage to the lateral control prevented it from turning. (*Coastal Command Review*)

This attack sank U-383.

The following day aircraft from the same two squadrons, DV968/M of 461 Squadron (Flt Lt I. Clarke) and JM708/N of 228 Squadron (Fg Off Hanbury) co-operated in another attack, resulting in the destruction of U-106. Another 228 Squadron aircraft had sighted three possible targets but on closer investigation these turned out to be torpedo boats (although reported by the crew as Z-Class destroyers). The aircraft shadowed the boats, sending out position reports – at which point it was joined by Fg Off Hanbury's aircraft, which shuttled between the enemy ships and the 40th Escort Group to direct the latter to the position. Meanwhile, DV968 and a 502 Squadron aircraft continued to shadow the torpedo boats. Flt Lt Clarke then spotted the damaged U-106 and in company with Fg Off Hanbury's Sunderland attacked the U-boat. The 461 Squadron crew was first to attack:

The sea was calm with a moderate swell. The U-boat took violent evasive action and tried to keep the aircraft on her beam. Considerable flak was experienced but the Sunderland's front gun kept spraying the enemy's decks and bodies fell into the sea; just before the depth charges attack the aircraft's front gun fire was so accurate that it prevented the relief crew from manning the U-boats's guns. Meanwhile, N/228 arrived on the scene and saw M/461's attack. This was delivered from the U-boat's port bow at 80 degrees to the track while the U-boat was still on the surface and turning to starboard. The depth charges straddled the target, the centre of the explosions being just abaft the conning tower. At the moment of the attack the U-boat was still turning to starboard and in doing so she presented her starboard quarter to the waiting N/228. Before the plumes of the first attack had subsided, N/228 ran in and dropped another seven depth charges. This attack was made only 30 seconds after the first and also resulted in a straddle. After the attacks the U-boat stopped and began to settle by the stern. Black smoke, white smoke and oil were pouring from her. The crew rushed out of the conning tower and began to jump into the sea. Some, however, attempted to man the guns again, but were mowed down by the concentrated fire from the two Sunderlands. Some forty minutes later the U-boat blew up and sank as the demolition charges went off. (*Coastal Command Review* II/4, August 1943)

The squadron lost four aircraft during this period, two on operations and two in accidents. During the same period the squadron was experimenting with a:

The synthetic trainer at Mount Batten for Sunderland crews in October 1943. A – plotting floor, B – instruments, C – clock, D – light trays, E – W/T base station, F – window, G – convoy at 100ft to 1in. Ken Delve Collection

***(Below)* The synthetic trainer: pilot table and W/T bench.** Ken Delve Collection

... two-gun belt-fed turret in place of the single magazine-fed one at present in the bow ... the turret as modified and fitted by 461 Squadron was unanimously agreed to be the best answer to the problem of improving the at present somewhat inadequate armament. Together with the Galley-gun devised by this unit and now fitted to all Sunderland aircraft in the Command, this now makes two successful and approved additions to the Sunderland's armament for which this Squadron is responsible. (461 Squadron ORB, August 1943)

4 August brought yet another success, DD859 (Fg Off A. Bishop) of 423 Squadron, on patrol in the North-West Passage, sinking U-489, despite receiving a hit from the U-boat's 4.7in gun! Having sighted a surfaced U-boat

> The Sunderland immediately lost height to deliver an attack, and had closed to within one mile when it became apparent that the U-boat was not attempting to dive. The captain therefore decided to manoeuvre for position in order to deliver an attack from dead ahead, but the U-boat circled in such a way as always to present her stern to the aircraft. Smoke puffs were seen coming from the U-boat's guns but no bursts were seen and no hits felt on the aircraft. When the captain decided that he could not outmanoeuvre the U-boat, he flew to a position about one mile up sun of it and turned in to attack at about 300ft. Fire was opened at the correct time – 1,200yd with the 0.5in guns in the nose – and an undulating approach made. The bullets were seen falling very close to the conning tower after the first few rounds. At 500yd the aircraft levelled out at 50ft and flew straight in, opening up with the 0.303 guns in the front turret. At that moment the aircraft began to be hit by the U-boat's flak.
>
> At 300yd the aircraft was seriously hit near the port wing root by a lucky shot from the U-boat's 4.7in gun. Although this damaged the aircraft's controls, and the flak increased in intensity, the captain pressed home the attack and released six depth charges. By this time a violent fire was raging in the wing root and the galley of the aircraft, and the ailerons and trimming-tab controls had become useless. The outer engine controls were also out of action, so the pilot switched off those engines and made a forced alighting. The aircraft bounced three times and the port wing dropped. The wing tip hit the water caused the aircraft to swerve violently to port and nose in. Six of the eleven members of the crew managed to get clear of the aircraft, which was then burning furiously. After about four minutes it sank. The U-boat approached the scene of the crash, and after fifteen minutes was within 200yd. The crew were standing on the forward deck taking to their raft. The U-boat continued to sink; scuttling charges were then fired and the U-boat disappeared. Twenty minutes later a destroyer picked up the survivors from the Sunderland, who were swimming in the water, and the Germans from the raft. (*Coastal Command Review* II/4, August 1943)

U-489 was a 1,688-ton Type XIV tanker; this valuable submarine was sunk on her first war cruise, having left Kiel on 22 July.

The failure of the group sailing policy and the losses suffered in recent weeks led Dönitz to recall six boats which had recently sailed and to cancel any further group sailings. Although he had to admit defeat under the present conditions, experiments in July with the *schnorkel* – which would allow U-boats to remain submerged for much longer periods than before – along with various other developments, showed some promise for the future; the U-boat campaign was by no means over. For Coastal Command, the Musketry and Seaslug areas dried up in terms of targets and were cancelled, being replaced by new operational areas, called 'Percussion' A to E, in late August, with additional Percussion areas being added later in the year.

There was still room for 'levity' even in official documents; the crew of a 461 Squadron aircraft had an 'interesting' encounter during a 3 August Musketry patrol:

> A raft, yellow in colour, was sighted and on investigation it was seen that there was a body on board, which, from its flat-white colour was deduced to be nude and, possibly, dead. The Captain of the aircraft dropped a marine marker at the spot and made a signal – 'Am over raft – one person aboard'. He circled around the raft and when about half-a-mile away from it saw the body dive overboard. When he passed over the raft for a second time it was empty. The obvious deduction was that the man on the raft was a Hun who did not want to be shot up by the Sunderland.
>
> Determined to save the diver from his own folly, the Captain of the aircraft went over to an Escort Group and informed the SNO that there was a 'raft with a man aboard'. A destroyer dutifully followed the aircraft to the raft. While engaged on this errand of mercy the Sunderland twice flew over the raft and each time the man was seen swimming alongside; but whenever the aircraft appeared, he went underneath the raft to escape attention. When the destroyer arrived it signalled 'no man aboard raft'. The aircraft replied 'man is underneath raft'. After again investigating the raft the destroyer flashed 'man on raft was a turtle'.
>
> These are the facts. It is not possible to doubt the word of the aircrew, who are all experienced and who of course were all sober at the time. Yet it must be admitted that the crew of the destroyer was in a better position to judge the species of the boy than was the crew of the Sunderland. This one might have been a tripper to the War Zone, who had become tired (or short of fuel) that he decided to rest on the raft. It is difficult to say, as he is not available for interrogation.

Many of the losses over Biscay were of unknown cause, although enemy action was no doubt the prime cause; however, on occasions the loss was both known and more unusual – as with 422 Squadron's DD861 on 3 September. The aircraft failed to return from a sortie over Biscay but the crew were eventually rescued and told their story. Whilst on patrol the starboard outer engine caught fire and dropped into the sea, taking the float and part of the wing with it. The starboard inner then failed because of the broken fuel line and the Sunderland was ditched; it stayed afloat for just two minutes but that was time enough for the crew to take to their dinghies. A pigeon was released (one of the few references to the actual use of pigeons, two of which were carried by each aircraft) and the two dinghies were tied together. The crew were eventually rescued on 6 September by a 228 Squadron aircraft.

8 October saw the destruction of U-610 by Fg Off A. Russell of 423 Squadron flying DD863. Sunderland JM712 (Fg Off A. Bellis) of 422 Squadron was escorting convoy ON206 on 17 October when it engaged a U-boat:

> Two radar blips were obtained at 5 miles' range. The aircraft was in a rain squall at the time, but on coming out of it the crew saw two surfaced U-boats 5 miles ahead. They were 740-tonners with a 37mm gun forward and two 20mm on the lower bandstand and two 20mm on the upper bandstand abaft the conning tower. There were also several machine-guns on the bridge. The position was about 20 miles south of the convoy and the sea was rough, with a north-westerly wind of 35kt. The Sunderland flew in to attack taking evasive action by undulating. The U-boat opened fire at 2,000yd with every gun that would bear. The aircraft's evasive action and the good shooting of the air gunners, who cleared the enemy's decks, enabled the Sunderland to escape damage on its first run. The attack was made from the enemy's port beam at 50ft but the stick undershot the target by 30ft. The Sunderland then did a tight turn to port at a range of ½ mile and came in at 100ft for a second attack. On this run no evasive action was taken and the aircraft was heavily shelled by both U-boats. The R/T was shot away and the front turret recuperator destroyed, the automatic pilot was blown out of the aircraft, the W/T destroyed and the radar damaged; the control quadrant was hit and the throttle and pitch controls shot away, the

Damage to DD865 after the 10 Squadron aircraft had been engaged by six Ju 88s over Biscay, 30 November 1943. Ken Delve Collection

wing dinghy was blown out, the mid-upper turret and the hull generally riddled. The navigator was mortally wounded but was able to give the pilot a course for the convoy before he died. The Group Gunnery Officer and the front gunner were also killed. In spite of this heavy and accurate shooting the captain carried on and from 50ft released two depth charges across the U-boat's beam. The third failed to release. The U-boat was straddled, lifted noticeably and disappeared without seeming to dive in the normal way. The second U-boat remained on the surface during throughout the action. After the action the Sunderland returned to the convoy, reported the attack to HMS *Drury* by V/S and informed the ship that it was going to ditch. She touched down on the top of the swell at 75kt about 100yd ahead of the escort. After bouncing once the aircraft buried its nose in the oncoming swell and disintegrated. The surviving crew expressed their great appreciation of the Navy's efforts in rescuing and caring for them. (*Coastal Command Review* II/7, November 1943)

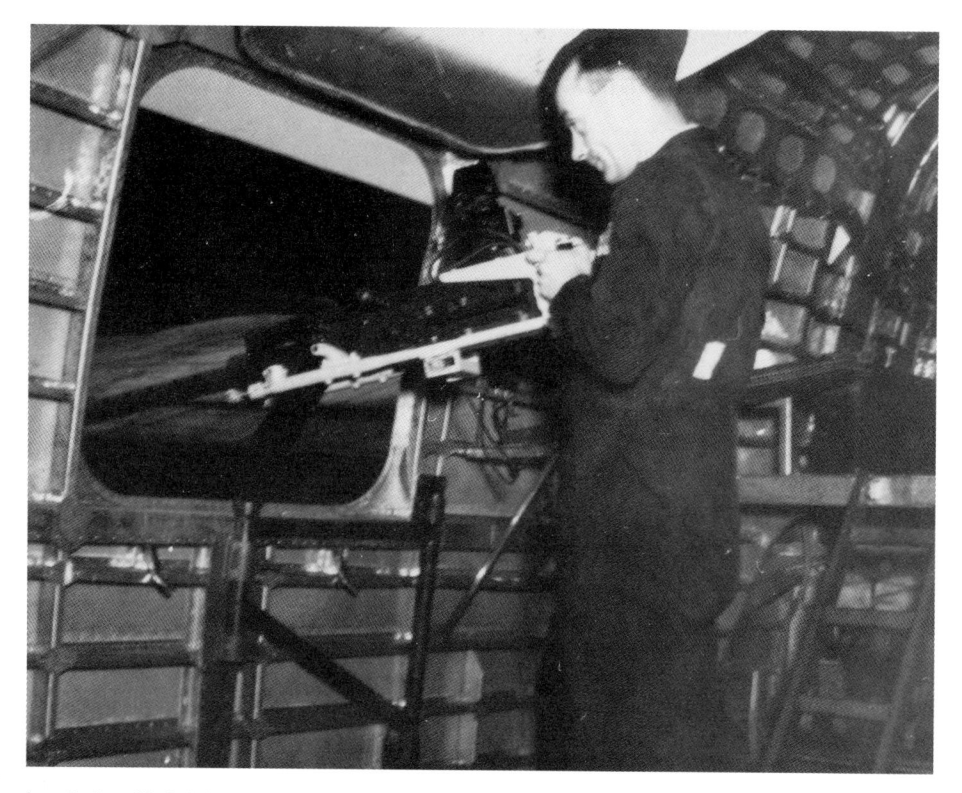

Installation of belt-fed 0.5in Brownings, February 1944, 10 Squadron RAAF.
Ken Delve Collection

The Sunderland units continued to lose aircraft and whilst it is not possible to record all such losses in this history, it is worth, as has been done throughout, 'flagging up' the occasional loss: No. 228 Squadron lost EK572 (Fg Off Franklin) during a Percussion patrol on 10 November, and DD864 (Flt Lt W. Fitzearle) on a Percussion T3 on 24 December; whilst 422 Squadron lost W6031 (Fg Off J. Ulrichsen) on 20 November. (All losses are detailed in the production list in Appendix I.)

There are many accounts of air combats involving German fighters engaging Sunderlands over the Bay of Biscay (some of which have been recounted above). The citation to Flt Lt Arthur Finucane's DFC read:

As captain of aircraft, Flt Lt Finucane has taken part in numerous operational sorties, invariably displaying skill and courage of a high order. In November 1943 during an ASP in the Bay of Biscay, this pilot's aircraft was attacked by four Ju 88s. A combat lasting forty-five minutes ensued during which the persistent attacks of the enemy aircraft were beaten off by the skilful airmanship of Flt Lt Finucane and the efficient coordination of his crew. Three of the enemy aircraft were damaged. In addition to the distinguished flying shown by this officer on that occasion and again later when he drove off a Fw 200 from the vicinity of a dinghy containing survivors, Flt Lt Finucane displayed exceptional skill in December 1943 when he flew his aircraft back having lost a propeller and use of the port engine. He made a masterly landing in the dark alongside a boat, with the aid of an Aldis lamp.

It was not only over the Bay that enemy aircraft were encountered. Although the Luftwaffe had few long-range aircraft, combats with Fw 200s and He 177s did occur from time to time as these aircraft either shadowed or attacked convoys. One such combat took place on 21 November when an aircraft of 201 Squadron engaged a number of enemy aircraft:

The Sunderland had just set course from the convoy when it sighted two enemy aircraft about 500yd away. The pilot altered course and increased speed in pursuit of the enemy, who were flying in and out of cloud. An He 177 appeared just ahead and above 'O', whose front gunner opened fire at 400yd, and saw tracer enter the enemy. The He 177 returned a few bursts and, although 8 miles away from the convoy, dropped a rocket bomb, which fell into the sea. Climbing to 3,000ft, 'O' saw two aircraft below, some miles away. He gave chase and, drawing nearer, identified them as Fw 200s. The front gunner opened fire on one at 600yd and saw tracer enter the enemy, who returned fire and increased speed. The Sunderland was forced to break off action owing to the enemy's superior speed. Later an Fw 200 dived out of cloud in front of the Sunderland and dropped a bomb. After a few more minutes the He 177 was sighted again; the pilot of 'O' attacked from 200ft below and on its starboard side. The front gunner opened fire followed by the port galley gun, and finally the rear gunner was able to bring his guns to bear also. When the enemy aircraft was out of range 'O' broke off the attack and, having reached PLE, left the convoy. A praiseworthy action by the captain of the Sunderland against three aircraft of superior performance and armament. (*Coastal Command Review* II/7, November 1943)

On 27 December Sunderlands were amongst a number of aircraft tasked with searching the Azores area for the blockade runner *Alsterufer*, the intention being for them to home surface vessels on to the target.

The enemy vessel was first sighted at 0845 by 201 Squadron's 'T' at position 4640N 1930W; the aircraft shadowed the vessel for four hours and then attacked with two 500lb MC bombs and two depth charges. The attack was made from 3,000ft above cloud and no results were observed.

First to sight the enemy was Flt Lt Baveystock of 201 Squadron at just before 1000, joined over the next few hours by two other Sunderlands whilst location messages were sent out to Group. A few hours later, with no sign of any surface vessels, Baveystock requested authority to make an attack. The high-level attack produced no result and, short of fuel, Baveystock left the area. A second attack was made by Sunderland EK576/Q of 422 Squadron, this aircraft failing to release its bombs on the first pass, although the gunners raked the decks of the ship. On its second pass, two 500lb MC bombs were dropped from 1,500ft but only one was seen to explode, and that 40yd short.

> Flt Lt W. Martin (EK576) was on anti-shipping search when he intercepted a sighting message from T/201. EK576 took up position opposite T/201 in circle, six miles from the MV at 1120. The aircraft attacked out of cloud at 1125 but overshot and did not release bombs. The second attack was made from 1,500ft at 1140, releasing two 500lb MC bombs, instantaneous fuze, 30 seconds space, at an angle of 20 degrees to the vessel's course. The decks were raked with machine-gun fire. One bomb was seen to explode 40yd undershot amidships, the action of the second bomb was unknown. A 4in shell struck the port wing after the attack was delivered. (422 Squadron ORB, December 1943)

A third Sunderland attack was made by 'U' of 201 Squadron, one 500lb bomb being dropped from above cloud with no result. The ship was subsequently critically damaged later the same afternoon in a rocket projectile attack by a Liberator from 311 Squadron.

In the latter part of 1943 Coastal Command had been considering the Martin Mariner as a possible replacement for the Sunderland; indeed, in September 1943 four of these aircraft had been delivered to 524 Squadron at Oban for evaluation. By December the aircraft had gone and the squadron disbanded, it being decided that the Mariner was inferior in endurance and carrying capacity to the Sunderland.

As the year closed, the Air Ministry was calling for even more trained crews: 'In view of increased demand for Sunderland crews beginning in March, it is imperative that the 4 OTU intake should be increased to maximum capacity.' (31 December signal from Air Ministry to Coastal Command.) The training problem was compounded early in 1944 by the introduction

8 January 1944 and a 500-ton U-boat (U-426) sinks by the stern, having been attacked by Fg Off Roberts of 10 Squadron RAAF. Ken Delve Collection

Fg Off Roberts and 10 Squadron RAAF crew of EK586 after sinking U-426 on 8 January 1944. Note the pigeon boxes being carried by the crewman second from left. Ken Delve Collection

of ASV III and the requirement by squadrons that crews arrived from the OTU 'proficient' in use of this equipment (it had been agreed in November to allocate the first three ASV III-equipped aircraft to the OTU, along with two experienced instructors from the ASV School at Chivenor). A conference was held at HQ 17 Group on 1 January 1944 to discuss OTU capacity: 'Can Alness cope with eighteen Sunderland crews at eighty-seven hours per month (i.e. 1,685 hours a month)?' Gp Capt Gordon stated that maintenance facilities were a problem and argued that the present maximum output was thirteen crews at seventy-two hours. It was stated that the ultimate requirement would be twenty-five crews at eighty-seven hours (total 2,400 hours) and Gp Capt Gordon was asked what was needed. He stated that a new slipway (plans for which had been prepared), more hardstands and additional maintenance equipment were essential. The introduction of the 87-hour syllabus was a result of the introduction of ASV III and over the next few months this subject was to feature again and again in the OTU policy files, including comments from squadrons that crews were not adequately trained in use of the equipment.

In January 1944, an advance party of 201 Squadron left Castle Archdale for Pembroke Dock – only to have the move promptly cancelled (although re-instated in March). The UK strength stood at seven Sunderland squadrons – one in the Shetlands at Sullom Voe (330 Squadron), three in Northern Ireland at Castle Archdale (201, 422 and 423 Squadrons), two at Pembroke Dock (228 and 461 Squadrons) and one at Mount Batten (10 Squadron RAAF). The battle against the U-boats, and especially operations over the Bay of Biscay, continued unabated.

January 1944 was a bad month for 228 Squadron in its operations over Biscay, three aircraft being lost. JM709 (Flt Lt Dowsett) failed to return on 6 January. JM708 (Flt Lt Grimshaw) was lost on 17 January when the starboard inner engine caught fire and the aircraft ditched, burnt out and sank. The crew made it into their dinghy and sister-aircraft 'S', circling overhead, directed three HSLs to the area. Finally, DW110 (Flt Lt Armstrong) failed to return on the last day of the month, having acknowledged an order from Group to divert to Castle Archdale but then not arriving on ETA. January also brought the first success for the year, U-426 falling to Fg Off J. Roberts (EK586) of 10 Squadron RAAF during a Percussion patrol.

U-426 was on only its second war cruise and had left Brest on 3 January for operations in the North Atlantic. The second success occurred to the west of Ireland when 461 Squadron's Flt Lt R. Lucas, flying cover for convoys SC151 and ON221, destroyed U-571.

461 Squadron's ML740 was shot down by Ju 88s of V/KG40 over Biscay on 23 March 1944, five of the crew being killed, the other seven being rescued.
Andy Thomas Collection

Riding the Waves

The Sunderland, like other flying boats, was not really at home on the open sea, a point stressed in the January 1944 edition of the *Coastal Command Review*:

> The rugged aspect of the Sunderland may perhaps mislead the landsman who can be excused for thinking that such a craft should surely be quite at home in the open sea. But for all its ark-like proportions the Sunderland and flying boats generally are mere eggshells of marine architecture. They are essentially aircraft which float and not boats which fly. For this reason their construction is as light as possible. The plates which form the hull of a flying boat are fashioned from duraluminium only $^{3}/_{32}$ of an inch thick and although considerable strength is achieved by frames and bulkheads, a flying boat does not compare with the smallest fishing smack as a sea-going craft.
>
> The weakest point in the seaworthiness of the flying boat rests in the wing tip floats. These are necessary to maintain stability. Without them the aircraft would roll on the water unchecked, until the whole boat capsized. The position of the float near the extremity of the wing does no lend itself to robust construction and the float is generally dependent upon struts and bracing wires. Thus, in the open sea, where large waves attack the aircraft from all directions, the wing tip float is subject to abnormal stresses. Should one carry away the flying boat will turn turtle and founder, even though the hull remains sound.
>
> In spite of their fragility, flying boats have been known to remain afloat for days in the open sea, given reasonably calm conditions. Flying

Trial installation of Pratt & Whitneys to a 10 Squadron RAAF aircraft, 27 March 1944. Ken Delve Collection

boats have successfully completed important missions which involved the hazard of alighting on the ocean swell. However, a recent analysis shows that for each landing and take-off carried out successfully, two have ended in disaster. Not unnaturally, therefore, Coastal Command has ruled that a flying boat may be allowed to land in the open sea only when special urgency, and the experience of the pilot, make the operation both necessary and advisable.

As noted at various places in this history, Sunderlands performed a number of notable rescues by taking such risks.

Another event of note in January was the departure of two Sunderlands (ML730 as A26-1 and ML732 as A26-3) from Mount Batten en route to Australia:

> The first two of six aircraft bought by Australia. On completion of petrol and oil installations by this Squadron the aircraft departed Mount Batten crewed by Tourex crews from 10 Squadron and 461 Squadron. (10 Squadron RAAF ORB)

The ORB also records an armament modification made during January:

> DD865 was modified to give it 'free' 0.5in Brownings on either side, just to the rear of the mid-upper turret. A stand for the gunners has been built in the aircraft. It has not yet been tested owing to a taxying accident. The Squadron is also having the nose turret replaced by an FN5 with two 0.303in Mk II Star guns.

On 15 February EK574/Q of 10 Squadron RAAF (Flt Lt J. McCulloch) was on patrol over the Bay of Biscay at 1,500ft when a formation of twelve Ju 88s was spotted about five miles distant:

> The captain of the Sunderland at once turned 180 degrees to port to reach cloud cover 7 miles away. When he came out of this turn he saw two other formations of four Ju 88s each, one of these formations, however, did not join in the attack. The twelve Ju 88s, now on 'Q's' port bow, split into three sections of four, and the individual aircraft formed line abreast whilst the formation of four on the starboard side went in to line astern. All the enemy aircraft were then flying at the same height as the Sunderland, which increased speed and held course until the enemy opened fire. The large formation attacked from 70 degrees on the port bow; at least eight of them and possibly the whole twelve, began firing simultaneously at 500yd and kept it up to within 100yd. 'Q' immediately made a climbing turn to port, and passed over the formation, which had apparently expected the Sunderland to dive as most of their fire went low. At the same time the four Ju 88s attacked from the starboard quarter, but were balked by the others and did not score any hits. During the attack the nose gunner fired at one of the aircraft in the leading formation and saw tracer enter the cockpit; he also got in a burst at the leading Ju of the first section and estimated hits. Only very slight damage was done to the Sunderland, but unfortunately a stray bullet killed the tail gunner outright. 'Q' turned to starboard again and reached cloud cover before a new attack could develop. Quick action on the part of the captain and good training all round undoubtedly saved the Sunderland and her crew from destruction by overwhelming odds. (*Coastal Command Review* III/2 February 1944)

En route to its patrol area on 10 March, Sunderland EK591/U of 422 Squadron (W/O Morton) spotted a U-boat on the surface (it was Morton's first operational sortie as captain and he was being screened by Flt Lt Butler):

> She was a standard 517-tonner without the forward gun, and she carried two twin 20mm mountings on the upper platform and one on the lower, the latter being well shielded. The Sunderland immediately manoeuvred to attack, reducing height to 400ft and the range to one mile. The enemy, however, sighted the aircraft as it turned and they opened fire at about 5 miles. Not unnaturally the fire was well short and it dropped about half way between the aircraft and the U-boat. The enemy also began to zigzag hard, but when the Sunderland was about a mile away the U-boat turned sharply to port and circled in order to keep stern on to the aircraft. Meanwhile, the Sunderland avoided flak by frequent alterations of height and course, and circled in an attempt to attack from the bow. It was soon evident that this was impossible and when the Sunderland was 1,000yd away on the U-boat's starboard beam, the pilot dived in to attack from the quarter. The last 400yd of the run-in was made at 50ft and the enemy put up intense flak, one shell hitting the aircraft's hull below the waterline. The Sunderland's front guns replied to such effect that in the end only one enemy gunner appeared to be serviceable. The attack was made from the U-boat's starboard quarter and six depth charges spaced at 60ft were released while the U-boat was still surfaced. The depth charges straddled the enemy's conning tower, one being seen to enter the water on the starboard side and three to port just abaft the conning tower. Three minutes after the attack the U-boat submerged but re-surfaced after another three minutes. She was then moving very slowly leaving a heavy oil trace and turning to starboard. An hour and a half later the U-boat signalled by V/S 'Fine bombfish', and the crew abandoned ship. Ten minutes later the U-boat sank by the stern leaving her crew in numerous small dinghies and one big one. The Sunderland crew were able to repair their damaged hull and the aircraft alighted safely at its base. (*Coastal Command Review* III/3, March 1944)

A Type VIIC, U-625 had departed Brest on 29 February on its second Atlantic cruise.

By March 1944 all serious opposition to the passage of convoys across the Atlantic had come to an end; between January and March, of 3,360 merchant ships in 105 convoys only three had been sunk, whereas the U-boats had lost twenty-nine of their number plus six more damaged. The intensity of the air effort, as well as strong and effective naval forces, had won this stage of the battle. The intensity of operations is reflected in many of the squadron record books:

U-625 goes down and the crew take to the water; the submarine was destroyed by W/O Morton of 422 Squadron on 10 March 1944. Ken Delve Collection

> During the month of March aircraft of No. 19 Group broke all previous records for flying hours, while it is admitted that the weather conditions were exceptional, this fact placed an even greater strain on all concerned and especially on the maintenance personnel.

Thus read a signal from Group recorded in the 228 Squadron ORB.

> It was forbidden to break radio silence in the vicinity of a convoy except for emergencies. Sitting above these eight ships with ocean spreading to infinity all around them, it was difficult to believe that they were in any kind of danger. The reality was, of course, very different. We were there to protect these vessels. It was clear from the reception we got that the escort ships valued the support that we were giving them.
>
> The most commonly used and indeed most effective method of sweeping an area of ocean in search of the enemy, or any other target, was the creeping line ahead (CLA) search. This entailed deciding the boundaries of the square or rectangular area that you wished to search and then covering the whole area with parallel sweeps from one side to the other. As many pairs of eyes as possible were brought to bear on scouring the area; the five main lookout posts in the aircraft were manned for the duration of the search. (Ken Robinson, *ibid*)

23 March was a trying day for 461 Squadron, two of its aircraft being attacked by enemy aircraft whilst on patrol over the Bay of Biscay.

> Sunderland F/461 encountered four Ju 88s; the enemy aircraft were first sighted 3 to 4 miles away on the port bow at 2,000ft and the Sunderland jettisoned its depth charges and lost height to 1,000ft. The 88s turned towards the Sunderland and flew up abreast, two on either side. Both aircraft on the port beam closed to

1,200yd, turned in to attack, and opened fire. 'F' turned sharply to port to meet this attack and began to corkscrew. The 88s broke off at 600yd and about 800ft above the Sunderland, the tail gunner of which fired a long burst at one of them as it passed astern. The two aircraft on the starboard side did not attack, so the fire controller instructed the captain to cease corkscrewing and resume course. The second attack opened in much the same way as the first, two 88s taking station on each beam. The starboard pair made a dummy attack but broke off at about 1,000ft above the Sunderland. Fortunately, the controller was not drawn by this manoeuvre, and the genuine attack from the port side was countered by a steep diving turn to port. By then the starboard outer engine had stopped, as a fuel supply line had been cut during the first attack. The Germans made four more similar attacks before they finally gave up the battle. The Sunderland captain took the same evasive action every time, ignoring the dummy attacks from starboard and countering the real attacks from port with steep diving turns into each attack. Only the first attack did any damage to 'F'. Excellent team work was chiefly responsible for the safe return of the aircraft from this combat. The navigator, who acted as fire controller, was always ready with the right instructions, the pilot handled the aircraft skilfully, and the gunners kept the enemy at a distance with accurate shooting. (*Coastal Command Review* III/4, April 1944)

In the other incident, aircraft Sunderland 'M' was engaged by nine Ju 88s, the enemy tactics being a similar series of attacks by pairs of aircraft, although in this case the genuine attacks came in from the starboard side. In a series of attacks various hits were made on the Sunderland and as the aircraft was getting ready to meet yet another attack the fire controller called that one of the port fuel tanks was on fire:

The captain decided that he would have to ditch and ordered everyone to take up ditching stations. All the turrets except one were vacated, but the mid-upper gunner continued to fire at the enemy until 'M' actually touched the water. The captain tried to land across wind along the swell, but there was too much drift. He therefore turned into wind and at the last moment turned back along the swell, thus cutting out most of the drift. The aircraft was thrown off the crest of the first wave, went down into the trough and bounced again. The captain tried to gain a little more speed, but the port inner motor had become unserviceable and, through lack of flying speed, the Sunderland hit the next wave half way up. The tail turret was knocked off and the aircraft sank within two minutes. One member of the crew went down with the aircraft and four of the others are missing. The first and second pilots were last seen clinging to one of the airscrews and are presumed to have gone down with the aircraft. The other two jumped into the water, but owing to the heavy seas and strong wind it was impossible to reach them with the dinghies and they were not seen again. The rest of the crew got away in two dinghies and were picked up fifty hours later by the destroyer HMS *Saladin*. During the hours of daylight on the 24th and 25th, aircraft kept constant cover over the dinghies. (*Coastal Command Review* III/4, April 1944)

At 1530 on 24 April DD862/A of 423 Squadron picked up a U-boat at position 5056N 1836W:

The U-boat's conning tower had two bandstands both of which contained guns. The captain [Flt Lt F. G. Fellows] held his course and height for about 8 miles and manoeuvred until he was 5 miles on the U-boat's beam. The U-boat turned hard to starboard in order to keep stern on to the aircraft and opened fire with medium flak which exploded with white puffs. The fire was accurate for line but fell short by 3 miles. The rate of fire and the fact that the bursts were in clusters of six or seven suggest that the enemy was using a gun of the Bofors type. When the aircraft was between the U-boat and the sun the latter slackened her rate of turn and the Sunderland began its run in. At 1,200yd the aircraft opened fire with the four fixed nose guns and the two front turret guns with such effect that for the last 300yd of the run the enemy's gunners did not reply. Up to this point the aircraft had been repeatedly hit but the captain pressed home his attack taking only the minimum of evasive action. He attacked from the U-boat's port bow and from 50ft released six depth charges spaced at 60ft. One depth charge exploded on impact with either the sea or the U-boat and the effect of this on the aircraft and crew prevented them

Another German submarine on the receiving end from a Sunderland; 21 May and U-995 was damaged by Plt Off King of 4 OTU. Ken Delve Collection

Sunderland V

The prime motivation for this variant was the need to find a replacement for the Pegasus engines, the superb Pratt & Whitney Twin Wasp being suggested by the experienced 10 Squadron RAAF. There was some suggestion that these engines would prove unsuitable without airframe modifications and in view of the need to increase rather than decrease production this would have proved unacceptable. However, Shorts looked at the problem and with Air Ministry support was contracted to modify ML765 to take the engines. At the same time a set of engines and technical drawings were sent to Mount Batten for 10 Squadron to convert ML839. Shorts flew ML764 in March 1944, ML839 following two months later.

The immediate advantage of the new aircraft was that a fully-laden Sunderland could now be flown with two engines on the same side feathered. Under the designation Sunderland V, the new variant was to enter production as and when Pratt & Whitneys were available, this taking place in the summer. First deliveries of the Sunderland V where to the Pembroke Dock squadrons, 228 and 461 Squadrons, in February 1945.

One distinguishing feature of the Sunderland V was the aerials of the ASV VIc, the scanners being positioned under the wing-tips. Some 150 Sunderland Vs were built, the last aircraft, SZ599, being launched at Belfast on 14 June 1946. A number of Mk IIIs were also converted to Mk V standard (*see* production list for details).

MAP Specification dated 18 November 1943

Powerplant:	Four Pratt & Whitney R1830-90B, S3C4G
Fuel/range:	2,137gal with 2,490-mile range, loiter endurance 17.9 hrs; 1,861gal with 1,960-mile range, loiter endurance 14.6hr; 2,458gal with 3,100-mile range, loiter endurance 22.8hr
Performance:	Service ceiling 19,700ft with max weight of 60,000lb. 22,600ft with mean weight of 51,900lb; max speed 229mph at 5,000ft, most economical speed 145mph, loiter endurance speed 132mph
Armament:	One 0.303in in nose turret, 700 rounds per gun four 0.303in in tail turret, 1,000 rounds per gun two 0.303in in side position, 800 rounds per gun 8 × 250lb, or 4 × 500lb, or 4 × 475lb DC

from seeing the explosions of the rest of the depth charges. As the rear gunner got the forward part of the U-boat in his sights there was a violent explosion which threw up everything movable in the aircraft – floorboards, IFF set and crockery; these, together with some eggs and the crew, formed a new variety of omelette on the edge of which the rear gunner lay unconscious. The wireless operator was thrown from his perch in the astrodome, all the electrical circuits became unserviceable, the R/T cable was cut, seams in the wings were opened and the port flaps were put out of action. The main damage, however, was to the elevator and all the strength and skill of the captain and second pilot was needed to overcome this. The aircraft began to climb and even with full nose trim the captain had to put pressure on the controls. Eventually all the crew were brought forward of the main spar in order to help maintain trim. When the aircraft was 300yd away the front gunner saw a brownish pool with blue smoke hanging over it 70–100ft astern of the U-boat. While the aircraft was being brought under control, the rear gunner, who had regained consciousness, saw the U-boat down by the stern and listing ... having climbed to 600ft and once more in control of the Sunderland, turned back over the target area and although there was a patch of oil there was no other evidence of damage to the U-boat. (*Coastal Command Review* III/5, May 1944)

Sunderland JM667/V of 330 Squadron took on U-240 on 16 May, the Sunderland suffering damage and casualties and the U-boat being sunk:

The Sunderland approached and opened fire at 1,000yd, the U-boat replying and hitting the aircraft on the bottom of the hull. At 800yd the front gunner opened fire and scored hits on the conning tower and gun crews. Both sides took evasive action, the aircraft jinking and the U-boat zigzagging and then turning violently to port.

The aircraft ran in over the enemy's port quarter at 50ft and attempted to release four depth charges, but all of them hung up. As the aircraft passed over the U-boat the rear gunner fired into the conning tower and several of the German gunners were seen to fall overboard. During the run-in the enemy's flak was very concentrated but after the aircraft had turned to port to attack again the flak was much lighter, only one gun being seen in action. The U-boat continued to circle to port while the Sunderland came in again from the port quarter. During this second attack the front gunner was killed, the third pilot and flight engineer wounded and the first pilot temporarily blinded by bursts of cannon fire in the cockpit as well as flames and oil from the front turret. Nevertheless, the pilot pressed home his attack and released four depth charges from 40ft while the U-boat was still on the surface. Three of the depth charge explosions were seen, the nearest being 15ft away from the U-boat's port quarter. Several of the Sunderland's crew saw the plume of this depth charge wash over the U-boat's stern. Meanwhile the aircraft had been hit again in the starboard outer engine, which stopped; at the same time the starboard inner engine began to vibrate violently. Immediately after the second attack, when the explosion plumes had subsided, the U-boat's bow submerged and then lifted clear out of the water at an angle of about 30 degrees. About three minutes after the attack the U-boat submerged stern first. The Sunderland, owing to its extensive damage and casualties, set course for base immediately. The captain had great difficulty in maintaining height and the crew were sent to ditching stations. After jettisoning 1,000gal of petrol, all spare ammunition, practice bombs and smoke floats, the captain was able to gain height and returned to base on three engines. (*Coastal Command Review* III/5, May 1944)

Two more U-boats were credited as damaged in late May, Plt Off E. King of 4 OTU hitting U-995 on 21 May and Flt Lt R. Nesbitt of 423 Squadron hitting U-921 on 24 May.

CHAPTER FOUR

D-Day and Afterwards

The D-Day Landings

Most anti-U-boat effort in late spring/ early summer 1944 was connected with the Allied build-up for and execution of the D-Day landings, an Admiralty assessment calculating that over twenty U-boats, in two main groups, were converging on the area in early June.

Flt Lt Bowrie of 228 Squadron (ML763) was in action on 5 June, obtaining two SE contacts and sighting a fully surfaced U-boat. The Sunderland tracked over the submarine at 75ft in what looked like a good attack – but the bomb doors failed to open. Having dropped a marker, the Sunderland positioned for a second attack but the U-boat was not re-sighted.

7 June was a busy night for the RAF's anti-submarine aircraft; it was also a successful night, a number of attacks being made, with two U-boats credited to Sunderland attacks, U-995 to Flt Lt L. Baveystock of 201 Squadron and U-970 to Flt Lt G. Lancaster of 228 Squadron.

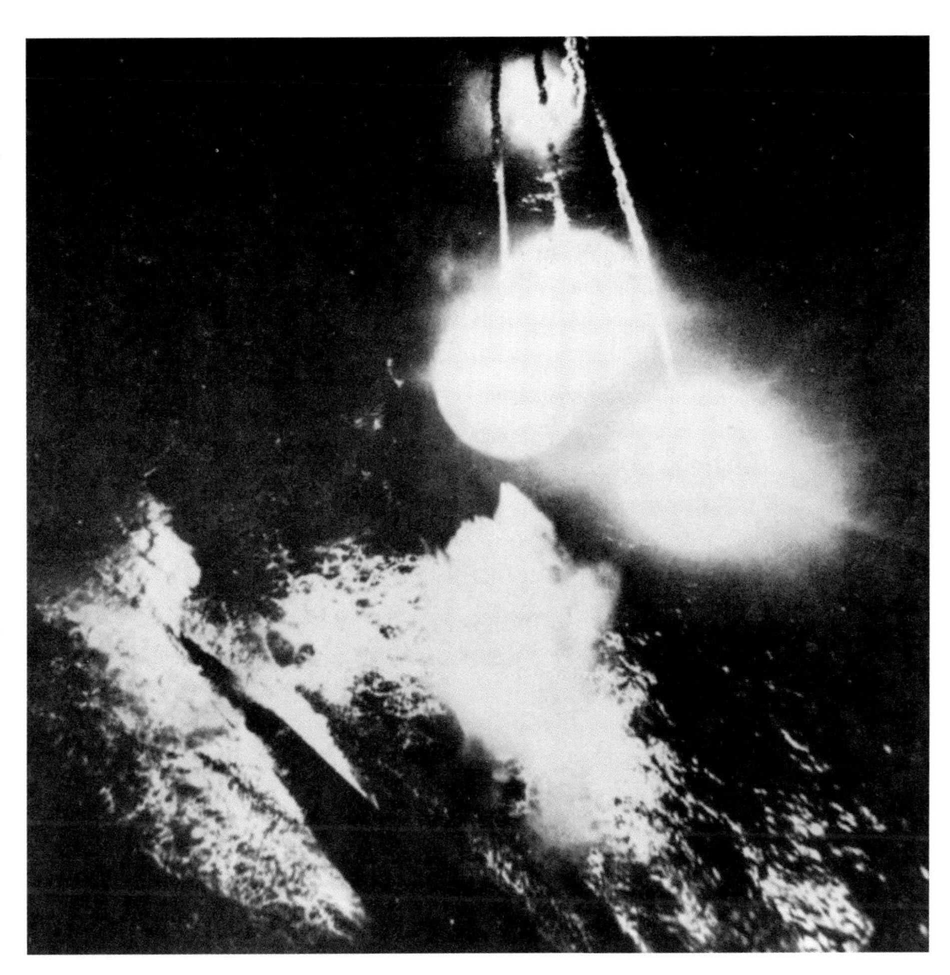

Excellent night attack; under the light of a flare ML760 of 201 Squadron, Flt Lt Baveystock, sinks U-955. Ken Delve Collection

> On the night of June 7, Sunderland S/201 [ML760] was flying at 450ft just below 10/10 cloud when a radar contact was obtained bearing Red 106 degrees, range 9 miles. The aircraft homed and the contact was down to half a mile, when the blip disappeared. The aircraft was then at 100ft and began to drop flares. Immediately afterwards the crew saw an unmistakable swirl and the second pilot saw bubbles rising to the surface. One minute later another Sunderland, N/461, dropped flares in the same area. The aircraft sent amplifying reports and began baiting tactics. Contacts which later proved to be from N/461 occasioned several false alarms, but some 2¼ hours after the original attempt 'S' obtained a firm contact bearing Red 70 degrees, range 11 miles. The captain immediately homed, the radar operator holding the contact throughout the run and giving faultless homing instructions. When the range was half a mile flak appeared dead ahead, but the tracer passed high and to port of the Sunderland. The aircraft turned slightly to starboard, dived and began dropping 1.7in flares which illuminated a U-boat 600yd fine on the port bow. She was steering due east at 10kt at position 4513N 0830W. At the moment of sighting she began to turn half to starboard and at the same time opened fire with her after guns, which was probably a quad mounting. The Sunderland turned in to attack and a six-second burst from the fixed and the turret guns silenced the enemy's fire so that the last 200yd of the run was unopposed. The Sunderland attacked over the enemy's port quarter at 75ft and released six depth charges spaced at 60ft while the target was fully illuminated by fifteen flares. The rear gunner saw the U-boat's silhouette completely blotted out by plumes. Three or four seconds after the depth charges had been released, and just before the explosions were seen, there was a heavy thud in the aircraft as if it had been hit. Throughout the attack flares were dropped and continuous radar watch was kept. The blip was held all through the run in but it disappeared immediately after the actual attack and was not picked up again. In spite of an hour's search the crew saw nothing more of the target. (*Coastal Command Review* III/6, June 1944)

A 210 Squadron Sunderland awaiting rescue; the aircraft had to ditch after the starboard inner propeller had flown off. Note the dinghies on the top of the wing. Ken Delve Collection

Shortly before midnight ML877/G of 228 Squadron (Flt Lt Lancaster) was on a moonlight patrol when the radar operator picked up a weak contact at 15 miles to starboard. The aircraft homed to the target area and dropped flares at half a mile:

> These revealed the U-boat bearing Green 20 degrees in position 4543N 0415W, steering west at 10–12kt. The Sunderland immediately altered course towards the enemy who, as soon as the first flare ignited, began firing wildly well to port of the aircraft. The Sunderland replied with the fixed guns and front turret which temporarily silenced the U-boat gunners; at 200yd, however, they again opened fire, this time accurately, and hit the aircraft in eight places. From 100ft the Sunderland released six depth charges at 55ft using the low-level bombsight. The bomb-aimer estimated that the centre of the stick overshot slightly but the mid-upper gunner, who was in the best position to see the results as he was not blinded by the flares, saw one small explosion plume subsiding on the enemy's starboard side and a big plume on the port side just abaft the conning tower. The U-boat's stern seemed to lift out of the water and she heeled to starboard. A big patch of oil 200ft across was then seen and emerging from this was a thin oily wake 400ft long at the end of which lay the U-boat, either stationary or stopping. Her gunners fired six rounds of tracer vertically in the air nowhere near the aircraft. The Sunderland stayed in the area for another 2 hours and 20 minutes but saw nothing more. (*Coastal Command Review* III/6, June 1944)

Other squadrons noted that 1.75in flare chutes were being installed in aircraft during the summer.

The final attack of the night was made by another 228 Squadron aircraft, ML766/R (Flt Lt J. Quinn) just after midnight:

> The radar operator picked up three contacts 5 miles ahead; soon afterwards the crew saw lights and then three vessels about 6 miles away up moon. Even with the help of binoculars it was not possible to say more than that the ships were escort vessels. A minute later another contact was picked up 8 miles away on the port bow. By using radar bearings the captain manoeuvred into a position down moon of the target and after 6 minutes he sighted a U-boat 5 miles away on the starboard bow at 4515N 0410W. Apart from the fact that she was probably a 740-tonner, no details of her superstructure could be seen, as dark patches of shadow on the water made observation difficult. Keeping the U-boat in his starboard bow the captain positioned the aircraft for an attack and used a patch of shadow in which to reduce his height to 300ft. At 1,000yd range he turned to starboard, the first pilot directing him towards the enemy's exact position. When the target was clearly visible in the moonpath the Sunderland ran in to attack. At 900yd the front gunner opened fire; the tracer however interfered with the pilot's vision and at 400yd the gunner was ordered to cease fire. The Sunderland left the shadow patch and from 50–70ft released six depth charges spaced at 55ft. The aircraft attacked over the U-boat's port bow and tracked over the target slightly forward of the conning tower. The enemy did not return the Sunderland's fire until just before the depth charges were released, which would seem to indicate that the attack was a complete surprise. The rear gunner saw the depth charges straddle the conning tower and he opened fire as the aircraft passed over. The explosions seemed to lift the enemy's stern out of the water and several of the Sunderland's crew saw the target enveloped in the plumes. The U-boat began to circle to starboard and then attempted to submerge. Her fore-casing disappeared but the conning tower and after-casing remained visible and, apparently, level. The U-boat continued to circle to starboard for 10 minutes, then steadied for one minute and began circling to port, losing way and settling lower in the water. The Sunderland dropped a marine marker one mile down moon and maintained visual contact with the enemy from that sector. The U-boat slowly lost speed and settled until about an hour and half after the attack only her conning tower was visible above water. Soon after this the radar contact, which had been getting smaller and smaller, faded altogether. (*Coastal Command Review* III/6, June 1944)

Date	Aircraft	Captain	Time	Details
1 July	DV980	Sqn Ldr Benton	0230–1541	ASP
	W6068	Flt Lt Lee	0144–1545	ASP
	EK581	Fg Off Pepper	0739–2121	ASP
	DP191	Fg Off Ulrich	1055–0004	ASP
2 July	DW111	Fg Off McCann	0310–1640	ASP
	DV978	Flt Lt Nesbitt	0913–2252	ASP
3 July	EK581	Fg Off R. Stillinger	0610–2005	ASP
	DD843	Plt Off Hemming	0850–2125	ASP
	W6011	Flt Lt Cunningham	1324–0212	ASP
4 July	DP193	Flt Lt Johnson	0408–1713	ASP
	DD849	Flt Lt Lee	0726–2055	ASP
	DW111	Flt Lt Heal	1313–2117	ASP
	DP867	Fg Off Campbell	1206–2153	ASP
5 July	DV980	Flt Lt Grant	0125–1406	ASP
	DP191	Fg Off Ulrich	0240–1545	ASP
	DV978	Flt Lt Lee	0619–1955	ASP
	W6068	Sqn Ldr Benton	1140–0105	ASP
	EK575	Flt Lt Fellows	1258–1218	ASP
6 July	DD867	Flt Lt Jackson	0049–1450	ASP
	DP193	Flt Lt Cunningham	0507–1822	ASP
	DD853	Sqn Ldr Grant	0550–1925	ASP
	DP849	Flt Lt Nesbitt	1125–0045	ASP
7 July	DP191	Plt Off D. Hemming	0220–1355	ASP
	W6068	Flt Lt Johnson	0248–1659	ASP
	DW111	Flt Lt Heal	0805–2225	ASP
	DV980	Flt Lt Lee	1252–0230	ASP

No credit was subsequently given.

The D-Day period saw most squadrons flying a record number of hours; typical was the 914 hours, in seventy-seven operational sorties, flown in June by 228 Squadron, during which four U-boat sightings were made, resulting in three attacks. The Squadron recorded that

> … an innovation was the flying of night patrols with the aid of flares, and three of the sightings and two of the attacks took place during the hours of darkness. The Squadron suffered the loss of two very good crews during these operations. Special patrols were laid on by Command to flood with aircraft the western end of the Channel and its approaches, and which entailed meticulous time-keeping on patrol and up to seven circuits of a comparatively small area. This was an added load on the shoulders of the navigators who carried out their work with commendable efficiency. (228 Squadron ORB June 1944)

After its early successes in the month, 228 lost two aircraft on consecutive days on Biscay patrols – ML762 (Flt Lt Hewitt) on 10 June and ML880 (Flt Lt Slaughter) on 11 June; the latter had fallen to the guns of U-333.

Another unsuccessful attack was made on 21 June when Plt Off Hart, in conjunction with a 502 Squadron Halifax, attacked a surfaced U-boat.

> During a recent attempt by the Germans to get their U-boats into the Atlantic via the coast of Norway, 4 (Coastal) OTU was called on at short notice to take part in the offensive against them. At the time of writing they had flown eight patrols, seven of which were completed – one aircraft having to return with engine problems. There were three sightings and three attacks, one U-boat was sunk and one was probably sunk. Any operational squadron would be proud of such results and for an OTU they are highly credible.
>
> Crews taking part were not picked men but simply average crews who were nearing the end of their course. Each aircraft carried an instructor to act as official observer and take over in case of emergency. (*Coastal Command Review* III/6, June 1944)

The summer months were busy ones for the Sunderland squadrons; June had seen many squadrons put in record levels of operational flying, many over 1,000 hours. This continued into July; 423 Squadron's sortie rate was typical (*see* table left).

The same level of operational flying – an average of four sorties a day – continued for the first three weeks of the month, the highlight coming on 12 July when Fg Off C. Ulrich (EK583) attacked a U-boat but with no apparent result. With the arrival of ASV III, the Squadron went into a two-week period of intensive training with the new equipment, along with training on the Low Level Mk III bombsight.

Meanwhile, July did bring some success for Coastal Command's Sunderlands. On 8 July three aircraft had a hand in the destruction of the U-243 at 4706N 0640W; first into action was Sunderland W4030/H of 10 Squadron RAAF, Fg Off W. Tilley and crew dropping six depth charges causing the U-boat's stern to lift out of the water. Sister aircraft JM684/K (Flt Lt R. Cargeeg) then delivered her attack, after which the U-boat appeared to be listing to port with white smoke coming out of the conning tower. The final attack was delivered about ten minutes later by a US Navy Liberator. The U-boat sank, leaving some forty men in the water; Sunderland 'H' dropped a dinghy, a Bircham barrel and a food pack; the crew were later picked up and in the interrogation related their view of the attack:

> We were making for Brest when at about 1420 hours on 8 July we came to the surface to fix our position. Immediately we surfaced we sighted two Sunderlands. The commanding officer decided to fight it out and we opened fire with the 37mm gun. One of the Sunderlands came in to attack from the starboard bow with cannon fire and dropped six depth charges. The last of these was a direct hit on the after part of the boat. As the aircraft flew off the rear gunner fired at us and put both the 37mm and the port twin 20mm guns out of action.
>
> The second Sunderland ran in from the port beam straight over the bridge, circled round the bows and came in again from the port bow firing cannon. Everyone on the bridge, including the commanding officer, was killed, except one petty officer and the second lieutenant, who was wounded. Others told me that this Sunderland dropped depth charges but everything was in such confusion that nobody knew just where they exploded. In the meantime the direct hit from the first aircraft had caused chaos below. The motors were on fire, the propeller shafts, propellers, hydroplanes and rudders were smashed, the diesels were torn from their mountings and everything in the engine rooms was badly knocked about. A diving tank was damaged and the W/T transformer was put out of action. The galley hatch was also forced open. Moreover, we had used up all the flak ammunition and it was therefore decided to abandon ship. Two seven-man rubber boats were cleared for taking off the ship's company but one was hit before it could be used.

Destruction of U-243 by W4030 *(top)* and JM684 *(bottom)* of 10 Squadron RAAF, 8 July 1944. Ken Delve Collection

The Liberator then made its attack but did no additional damage, after which the Sunderland '... threw us dinghies and supplies and later two Mosquitoes circled us until the arrival of the rescue ships some seven hours after we were sunk'. (*Coastal Command Review* III/7, July 1944)

By the end of June 201 Squadron had received a full complement of Sunderland IIIAs, the improvements on which included better armament (fourteen guns), plus

ASV III and Gee II. The new aircraft soon scored a success when on 11 July U-1222 was sunk by Flt Lt I. Walters and crew in ML881/P. The aircraft was on patrol in good weather when

> ... the second pilot sighted a wake about 8½ miles away on the starboard bow. No initial radar contact was made and the set was switched off 30 seconds later. The aircraft dived towards the wake and the crew clearly identified the periscope and raised schnorkel of the U-boat, the sighting was first made by the use of binoculars. The U-boat held its course until the aircraft was about ¼ mile away and then she seemed to stop. The aircraft was at a height of 100ft and the captain could then clearly see the protruding stern of the U-boat, which was apparently beginning to dive. The Sunderland attacked from the enemy's port quarter and from 50ft released five depth charges spaced at 60ft while the schnorkel, periscope and disappearing stern of the target was still visible. When the Sunderland returned to the scene of the attack the crew saw nothing but a few pieces of wood and whitish froth on the surface of the sea. About 4 hours later a 461 Squadron Sunderland scoured the scene and noticed a fresh red patch on the surface of the sea, about 50yd in diameter. Photograph evidence was used to assess that the No. 2 DC had proved lethal. (*Coastal Command Review* III/7, July 1944)

This was one of the few Type IX boats sunk by Sunderlands, the vessel was returning from its first war cruise, having been operating off the Canadian seaboard.

Early August saw the anti-submarine squadrons flying patrols at the west end of the Channel as a blockade to prevent U-boats from reaching the all-important resupply convoys crossing from England to the invasion area around Cherbourg. However, as this threat diminished the patrols were moved west of Brest. When the bases at Brest, Lorient and St Nazaire were no longer usable, the submarine hunters transferred their attention to an area nearer Bordeaux, just west of the Gironde.

By August Coastal Command was instructing 4 OTU to cut down the output of crews, the shortage at the beginning of the year having become a surplus. However, August brought a number of attacks and three confirmed U-boat kills, two to 461 Squadron (11 and 12 August) and one to 201 Squadron (19 August), with a number of other attacks having been carried out without definite result. Fg Off G. Bunting and crew in ML789/M of 228 Squadron attacked a U-boat on 9 August. Two depth charges were dropped in the first attack against a surfacing U-boat, but with no result. Ten minutes later the crew spotted a periscope and attacked with six DCs from 80ft: 'Heavy black oil welled to the surface and formed a large oil patch. About an hour later violent air bubbles were seen followed by filmy oil.'

Two days later Plt Off I. Southall in ML741/P of 461 Squadron was involved in the destruction of U-385, a Type VIIC boat on only its second war cruise:

> The Sunderland captain turned away until the range had opened to 8 miles and then turned back towards the target, which had been held all the time by radar. He made a series of S turns to get the U-boat in the moonpath, eventually sighting her again at a range of 6 miles. At 4 miles he lost height to 500ft and at 2 miles he dived to attack, releasing from 100ft six depth charges spaced at 55ft. He used no illuminant, as he could see the target up moon in perfect visibility. The depth charges were dropped across the U-boat's beam, and are stated to have straddled her amidships, four entering the water to starboard and two to port. The explosion plumes completely obliterated the target. Immediately before the attack the U-boat fired a few shots and what may have been a red recognition cartridge. At the moment of the depth charge explosions a bluish-white flash was seen from the U-boat. After the plumes had subsided she was still on the surface, but was stationary. The Sunderland turned to port after the attack and flew past the enemy's stern at a range of 1,000yd.

Castle Archdale was home to the Canadian Sunderland units; this is 423 Squadron's EK583 on the slip in July 1944. Andy Thomas Collection

The German gunners opened up with about six guns, but their fire, though fairly heavy, was most inaccurate. At 2 miles range the U-boat disappeared from view, and two minutes later the blip also disappeared. The Sunderland flew off to make contact with an Escort Group, which was then 9 miles SE of the position of the attack, and led the ships back to the area. On returning to the marker the crew found a patch of oil 100yd wide. Later they picked up a radar contact at 3 miles on the U-boat's original track about 2 miles ahead of her last position. After three attempts this was illuminated with flares and identified as a radar decoy balloon. The Sunderland reported the position to the ships and continued to stand by. About two and a half hours after the attack the radar operator picked up another contact, but it disappeared when the aircraft was a mile away. This was also reported to the ships, which began to sweep towards the new contact. After about an hour they reported having found an empty dinghy. After having stayed in the area for three and a half hours the Sunderland reached PLE and set course for base. At 0636 on August 12 the U-boat surfaced 3,000yd ahead of HMS Starling and was heavily engaged and hit by gunfire. She sank five minutes later. (*Coastal Command Review* III/8, August 1944)

The Squadron notched up another success a few days later, Fg Off D. Little and crew in ML735 sinking U-270:

The U-boat, which was steering 110 degrees at 15kt, looked very big and had a high conning tower with two bandstands aft. About ten men were seen in the conning tower or manning the guns aft of it. These gunners promptly opened fire with light and medium tracer which came up in four streams from points abaft the conning tower. The Sunderland replied with 120 rounds from the nose gun and 380 from the fixed guns. This fire discouraged the U-boat's gunners so effectively that the last 400yd of the Sunderland's approach was unopposed. The aircraft attacked from abaft the U-boat's starboard beam and from 300ft released six depth charges spaced at 55ft. The aircrew did not see the points of entry of the depth charges and saw only two of the explosions, the nearer being about 15ft from the port side of the conning tower. At 2 miles' range the contact disappeared and when the Sunderland returned to the position ten minutes later, intending to attack again, the crew neither saw the U-boat nor picked up any contact. However, about an hour after the attack, they saw many small lights in the water, two of which were flashing SOS. About 0130 hours some escort vessels arrived in response to the Sunderland's homing signals and 20 minutes later the SNO reported by R/T that the U-boat had been sunk and survivors rescued.

The engine could be turned over using a crank inserted into the cowling.
Arthur Banks

From interrogation of the crew it was discovered that U-270 had a scratch crew on board and had been trying to sail to safety to the south with an escort of three M-class minesweepers. The escort was recalled late on August 10 and the U-boat was soon having to react to air attack warnings, having to dive on a number of occasions whilst trying to reach La Pallice. At 0100 on August 13 the Sunderland made its attack and whilst none of the depth charges hit the U-boat they caused damage to the pressure hull near the No. 5 diving tank. The vessel attempted to continue on the surface but water was entering and she was in danger of sinking. Some time later a Wellington flew over the U-boat and the German crew abandoned ship before an attack was delivered.

The aim of the attack is to place a stick of depth charges or bombs so that it straddles the conning tower. In order to reduce errors to a minimum the actual attack is normally carried out from as low an altitude as possible. If eight weapons are carried and a bombsight is used, drop six in the first attack, followed by two in the second attack, if there is no bombsight drop all eight in one attack. Spacing should be 60ft for six or more bombs dropped without a bombsight, 100ft for five or less bombs. Spacing should be 45ft for eight bombs dropped with bombsight, 55ft if six bombs. An attack should be delivered with depth charges on any U-boat that has been submerged for not more than 30 seconds. Depth charge settings should be 25ft. The hull splitting radius is 19ft for a depth charge, the severe damage radius is 38ft. (Air Ministry instruction August 1944)

Concern over the 'next generation' of flying boats continued and on 7 August 1943, Coastal Command had written to the Air Ministry Director of Policy:

I don't know if the Sunderland IV is going to do what is claimed for it, but if it does it will be a very useful anti-submarine aircraft for Atlantic and European waters. But I gather that Shorts are not really getting a drive on with the Shetland; and I am afraid if we are not very careful we are going to be caught with the war ending in Europe and a real need for a Long Range aircraft in the Far East which we shan't be able to meet.

The *Coastal Command Review* contained an account of an attack carried out by a 201 Squadron aircraft on 18 August 1944:

The second pilot was flying the aircraft while the captain was on the lower deck. Simultaneously the front gunner, who was using binoculars, and the second pilot reported a wake four miles away on the port bow. The aircraft continued on course until the wake was on the port beam, when it was seen that the wake was caused by a periscope moving SE at four knots. The Sunderland captain took over in the second pilot's seat and attacked the periscope from Red 90, releasing from 50ft six depth charges spaced at 50ft. The depth charges straddled the periscope and exploded as the aircraft was turning away to port. Immediately after the explosion plumes had subsided, a great circular eruption of bubbles rose slightly above the surface of the sea and continued to effervesce for 20 minutes. A few minutes after the attack oil came to the surface and an hour and a half later the slick had grown to a length of 2 miles. Miscellaneous wreckage also appeared, including pieces of wood and sheets of white paper. An escort group and two Wellingtons were homed to the area before the Sunderland left. The surface vessels picked up some of the wreckage, identifying the white paper as German charts.

The Coastal Command writer commented:

> A very fine look-out enabled the captain to deliver an exceedingly good attack. The crew are to be congratulated on their excellent drill. It is of great value to be able to publish the fact that a U-boat showing only her periscope can be killed provided that the attack is accurate.

Flt Lt Baveystock and crew, flying EJ150, were credited with the destruction of the U-107. Although Baveystock was awarded an immediate DSO for this action, his comment was:

> My second pilot and the crew did it all for me. I was just buttoning up my trousers (having been in the lavatory) when I heard the alarm; when I got to the bridge they said, 'hey, look, there's a periscope' and I looked and there was a periscope, so I just sat in the second pilot's seat and pressed the tit. It was easy; I'm going to do the rest of my ops blindfold! (From *On the Step*, *ibid*)

In the evening of August 22 Sunderland A/228 was circling a marker dropped by another aircraft when the crew sighted a flash of light about 9 miles to the north-west. The captain turned towards the position, and soon afterwards the radar operator got a contact 2 miles away on the port bow. The aircraft homed and the crew sighted a periscope and schnorkel half a mile away on the port side, position N4621 W00146. The schnorkel was moving westwards at 2kt. As the Sunderland turned away to starboard to get into position, the mid-upper gunner opened fire. The Sunderland then turned steeply to port and attacked from Red 30 to the enemy's course. The captain released six depth charges spaced at 60ft, while the schnorkel and the periscope were visible. The second pilot, who was at the port galley hatch, saw the schnorkel in the middle of the splashes of entry before the explosion plumes obscured it. Three minutes later a twisted object came up in the middle of the depth charge scum. It was about 8ft across, looked like an aircraft fin much distorted, and was apparently attached to a larger body which could be seen just below the surface of the sea. The captain then turned and attacked this wreckage with one depth charge. After the explosion the wreckage was still visible and a small oil slick formed to the south-east of it. The Sunderland captain then returned to the marker which he had been circling at the beginning. However, he found another aircraft there and so he returned to the scene of the attacks. The depth charge scum was still visible and a short distance away were six or seven patches of heavy black liquid, each patch was about 30ft across. The aircraft then attacked the wreckage with the one remaining depth charge. Despite this damage the object continued to give a firm radar contact. (*Coastal Command Review*)

As Allied ground forces continued their advance through France, September 1944 saw an attempt by the U-boats to flee the bases in France and make for Norway; Nos 15 and 19 Groups were given additional air units to help counter this move. Attacks were concentrated along the coast of Norway to catch the submarines in transit, although much of this work fell to Coastal Command's Strike Wings rather than the maritime patrol aircraft.

Whilst Leigh Light Wellingtons had proved very successful in the Biscay campaign, effective night illumination was a continued problem for the Sunderland, the 1.75in flare remaining the main option. From mid-September, 423 Squadron was employing the flare in a more unusual way:

> A flare drop aircraft has been sent out each night over an area to the north where U-boats are suspected to be on the prowl. Starting at 2000 hours the aircraft drops a flare every four minutes to 0630 hours, from 1500ft. No sightings have yet been made nor has there been any evidence of the effect which this has on the U-boats (423 Squadron ORB)

The month brought varying fortunes for the Canadian unit, one U-boat being attacked on 11 September by Fg Off Farren.

Almost all sightings and attacks at this period were on schnorkelling U-boats – gone were the days of finding German submarines on the surface. The problems of locating, by radar and visually, the wakes left by the schnorkel or periscope, taxed the flying boat crews and it became increasingly important to have well-trained aircrew. Visual pick-ups were particularly important:

> On September 11 Sunderland D/423 was patrolling at 1,200ft over a calm sea when the crew sighted a thin whitish vapour or steam on the surface of the sea some 9 miles away. The radar was on but no contact was picked up. The captain turned to investigate and lost height. With the aid of binoculars it was seen that the vapour was issuing in a low-lying stream from a point on the surface. When the Sunderland was still some 2 miles off the vapour ceased as if it had been cut off but then a slight wake was seen extending for about 100ft. The Sunderland

Leslie Harold Baveystock

Les Baveystock was one of the most successful Sunderland captains, being credited with two confirmed U-boat kills plus a number of other attacks, scored with 201 Squadron in 1944.

Les Baveystock was born in Finchley in 1914 and enlisted in 1940, undertaking flying training in Canada before joining 50 Squadron. He was awarded a DFM, subsequently converting to flying boats and joining 201 Squadron.

Air Ministry Bulletin 12702 included notification of a DFC for Les Baveystock:

> This officer was the pilot and captain of an aircraft which sighted an enemy blockade runner on December 27, 1943. After signalling the position, Flying Officer Baveystock determined to attack the vessel. In the face of considerable anti-aircraft fire he raked the ship with machine-gun bullets and then attacked it with bombs which he released on his third run over the objective. His aircraft had been hit but he flew safely back to base. Visibility was extremely poor and the flare path could not be seen; nevertheless, in absolute darkness, Flying Officer Baveystock brought his aircraft down on to the water close to the shore with masterly skill; no further damage was sustained. This officer displayed outstanding keenness, efficiency and determination.

In August 1944 AMB 15346 noted the award of a Bar to the DFC but gave no citation details. Finally, in October 1944 AMB 15916 noted the award of a DSO to acting Flight Lieutenant Baveystock:

> Flight Lieutenant Baveystock continues to set the highest example of skill and leadership. His record as captain of aircraft is one of consistent efficiency and devotion to duty, and his work has contributed much to the success of operations against the enemy surface and underwater craft.

Sunderland IV

Under Specification R8/42 an improved Sunderland was proposed by using the Bristol Hercules engine installation from the Short Stirling. This would involve a heavier airframe and changes to the planing bottom. An order was placed for two prototypes, MZ269 and MZ271, and thirty production aircraft; design work concentrated on the planing bottom, whilst keeping as close as possible to the original basic design in order to ease actual production. Overall hull length was increased by 39in but one of the main visual differences was the deeper rear step.

A 5 degree dihedral on the tailplane was employed to increase spray clearance.

Overall empty weight rose to 19,000lb, part of this being in heavier armament, including a Bristol B17 mid-upper turret with twin 20mm cannon, an FN83 nose turret with twin 0.5in Brownings, and a similarly-equipped Glen Martin tail turret. The aircraft was powered by four 1,700hp Bristol Hercules XIXs, driving 12ft 9in fully-feathering four-bladed DH propellers.

Prototype MZ269 first flew at Rochester on August 30, 1944; however, early tests showed that there was a problem with lateral control. The fin was redesigned – and made 33in taller – and the tailplane area was increased by 20 per cent. The

attacked down track from ahead and from 50ft released four depth charges spaced at 70ft. He aimed at a point about 700ft ahead of the apex of the wake. After releasing the depth charges the Sunderland dropped a marker and circled the position. Some three hours later escort vessels arrived in response to homing transmissions. They picked up a positive submarine contact and made a depth charge attack. The Sunderland saw no evidence of damage to the U-boat by either air or surface attack although it was over the position for nearly six and a half hours. (*Coastal Command Review*)

navigation fit of its aircraft with the installation of Gee or Loran, although the latter presented a few problems in that the electrics of the aircraft had to be modified to provide adequate power supplies. Furthermore, this unit, like others, was finding that ASV III required more training for the operators and

> ... in an effort to increase the operational efficiency of the Squadron to combat the present advantage the U-boat has gained with schnorkelling, an intensive training drive is underway, to be completed by December 1. Two submarines, HMS *Vulpine* and HMS *Upshot* have been provided by the Navy to increase the efficiency of radar operators.

Departure of the Allied Control Commission for Norway from Woodhaven aboard a 330 Squadron aircraft. Coastal Command Records

Whilst Coastal Command assessed that the air attack probably failed to cause any serious damage it was considered that damage may have been caused to the delicate schnorkel system and that this may have caused the submarine problems later – perhaps causing it to surface and thus be more vulnerable to attack. On the down side, ML823 (Fg Off McCann) had failed to return on 6 September, only one survivor being picked up.

The following month the squadron was having improvements made to the

A few weeks later, on 6 November, the Squadron recorded that

> During the present lull in U-boat operations, more intensive training has been underway in radar homing, night bombing, use of the Mk III low level sight with 1.75in flare for illumination of the target. ASR launches fitted with radar

final change to ease this problem was a dorsal extension to the fin leading edge. However, with the Sunderland V planned to enter service at around the same period, the decision was taken not to proceed with this variant – at least not as the Sunderland IV; the first eight aircraft of the production batch were completed and delivered, as the S45 Short Seaford, for operational evaluation by 201 Squadron in early 1946. This was not considered to be a success and the RAF was thus not interested in adopting the type. One aircraft, NJ201, had been evaluated by Transport Command, again with no positive result; after modification to the nose and tail for streamlining this aircraft became G-AGWU with BOAC for evaluation as a civil transport. Six of the production Seafords subsequently went to Belfast to be converted into Short Solents for BOAC, in order to join the existing Solents operated by the airline.

MAP Specification dated March 10, 1945

Powerplant:	Four Hercules XVII
Operational:	2,835 gal for 2,480 mile range
Max bombs:	2,354 gal for 2,090 mile range
Max fuel:	3,065 gal for 2,725 mile range
Service ceiling:	10,500ft at max weight of 75,000lb.
Service ceiling:	15,000ft at mean weight of 63,930lb.
Max speed:	245 mph at 2,000ft, most economical speed 163 mph. 8 minutes to 5,000ft.
Armament:	Four 0.303in forward body, 500 rounds per gun
	two 0.5in in nose turret, 500 rounds per gun
	two 0.5in in tail turret, 1,000 rounds per gun
	two 0.5in in top body turret, 600 rounds per gun
	two 0.5in in side positions, 400 rounds per gun
	4 × Mk VII DC, total 1,680lb

(Top) **Sunderland IV prototype MZ269 at Rochester; the type entered service, but only for evaluation by 201 Squadron, as the Seaford.** Peter Green Collection

(Bottom) **Prototype Shetland I DX166, May 1945. This 'follow on' Sunderland came to nothing.** FlyPast Collection

reflectors were used as targets in Lough Neagh, whilst Innishmurray Island in Donegal Bay was also used as a target.

A number of convoys were re-routed to the south of Ireland rather than the North Channel and, to cover this, No. 15 Group flew patrols over the area, additional aircraft being provided by various detachments. There was, however, a general lull in the U-boat war after the hectic activity of spring and summer. On 3 November 201 Squadron returned to Castle Archdale and resumed Atlantic patrols.

'The mid-upper gunner sighted grey smoke and the aircraft at once turned to investigate, action stations being sounded,' stated a 201 Squadron report for 16 November. Flt Lt Hatton and crew in 'Y' were on an anti-submarine patrol:

> As the aircraft closed the smoke was seen moving at a considerable speed, estimated to be 15kt. Just before the aircraft tracked over the smoke, the front gunner distinguished a moving wake with the smoke issuing from the apex. The wake was approximately 2yd wide and nearly 100yd long. At the apex was a small horizontal wave, or splash, rising 2ft. This possibly screened the object making the wake. There were also distinct bow waves on either side of the apex and greyish bubbles were seen in the middle of the wake. The mid-upper gunner saw a narrow cigar-shaped, brownish shadow or shape beneath the surface of the sea. This sighting was confirmed by the captain. The aircraft dropped a flame float and began a climbing turn to port, circling its marker twice to port. During another turn the smoke and wake were seen again some 4 miles distant. The aircraft immediately lost height and made an attack down sun, crossing just ahead of the wake and releasing, from 50ft, six depth charges set to shallow depth, spaced at 60ft. During the attack the wake remained visible. When the early depth charge explosions were reaching their maximum height the rear gunner estimated that four depth charges undershot and two overshot. The aircraft searched the area for some time without seeing anything; four escort vessels were subsequently homed to the area and a Liberator took up station.
>
> The Coastal Command analysis, having checked the photographs, was of the opinion that this was a schnorkelling U-boat and that the depth charge fell just outside of damage range from the target, being some 150–200ft astern of the head of the wake. (*Coastal Command Review* III/12, December 1944)

J/10 Squadron RAAF on the rocks having broken its moorings, 2 September 1944.
Ken Delve Collection

In the middle of December a new German U-boat campaign commenced, significant numbers of boats operating in the Channel and inshore waters of south-west England and Ireland: 'The skilful use of schnorkels made them difficult targets to locate, although no attacks were made during the month, a considerable number of radar contacts were reported, many of them promising.' (228 Squadron ORB, December 1944). The difficulty of locating schnorkelling submarines made crews more observant of 'underwater disturbances' and a number of attacks are recorded on such phenomena, the assumption being that the disturbance might be caused by a submerged U-boat.

The next success, albeit not credited until an April 1945 Admiralty assessment, went to Flt Lt D. Hatton of 201 Squadron when, on 6 December, Sunderland 'Y' sank U-297 off the Hebrides.

Changes continued to be made to the training system; in January 1945, No. 4 OTU was allocated six of the Sunderland V conversions (with ASV III) as well as six new-build Sunderland Vs (with ASV VI). The same month it was decided that all Sunderland training at 131 OTU would transfer to 4 OTU.

Anti-submarine operations continued in the Channel and inshore waters into early 1945, although patrols were also flown over the Irish Sea. Despite more experience amongst the radar operators, locating U-boats remained a problem and although squadrons recorded a significant number of attacks on 'possible targets' (423 Squadron, for example, noting four U-boat attacks in January), few of these were proved to be definite U-boats and no damage/destroyed credits were given. The introduction of ASV VI-equipped Sunderland Vs in the early part of the year held promise of better results.

On 27 April Flt Lt Chapman (RN283, 228 Squadron) attacked the wake of a possible periscope, dropping six depth charges; unfortunately, the Sunderland was damaged by the explosions and crashed into the water. Seven of the crew managed to escape in a dinghy and were lucky that a Sunderland of 461 Squadron was able to land and pick them up.

When Flt Lt Ken Foster and crew of 201 Squadron's ML783 shared in the destruction of the U-242 on 30 April it was not only the unit's final kill of the war – indeed, the last Sunderland 'kill' of the war – it was also the first time that they had successfully employed 'High Tea', the code-name for an early version of the sonobouy. Ken Foster later said:

> After we got the contact at 0810, the bomb doors failed to open and we had to attack with the fixed and turret machine-guns. We signaled Group that contact was lost – nothing. At 0915 we asked for instructions – again nothing. At 1000 we sent 'resuming patrol'. When we regained the contact at 1135 we hand-lowered the bomb racks and straddled the target's course with our attack. We sent further signals to Group and again asked for instructions with again no reply. At 1235 we got the message 'comply with my 1135'. What flaming signal? I think it turned out that my WOP just received that signal as we were banging out a sighting report at 1135, put it under his folder and forgot about it. When we got home it was rockets all round, not a word of congratulations for the first two submarine attacks for months, although the Navy were very good and send us a note and a copy of the ASDIC trace. (From *On the Step, ibid*)

Admiral Dönitz – who by now had succeeded Hitler as *Führer* of Germany – issued an order to his U-boat crews on 4 May to cease hostilities and surrender; Coastal Command aircraft were ordered to – warily – shepherd surfaced U-boats to Allied-held ports. Many of the squadrons undertook such work; on 9 May Flt Lt Arigihi of 201 Squadron escorted U-1105 to Loch Eriboll. Not all German submarines obeyed the 'surface and surrender' instruction, and two days later another 201 Squadron crew (Flt Lt Thomas) dropped sonobuoys and depth charges on a submerged U-boat in the North Channel.

The last Coastal Command operational patrol of the war was flown by 201 Squadron Sunderland ZM-Z (Wg Cdr Barrett) on 3 June, an uneventful convoy patrol. New tasks were soon to be found for many of the units, but for others the end of the war meant disbandment. One of the first Sunderland units to disband was 228 Squadron; having operated Sunderlands since November 1938 and seen action in the Middle East, West Africa and the UK, the unit bade farewell to its flying boats on 4 June, very much the end of an era (the Squadron reformed the following June as a Liberator unit, still in the maritime role).

In May a number of Sunderland units had been told that they were being transferred to Transport Command with effect from early June, the need for 'transport capable' aircraft in the immediate post-war period far outweighing available resources; indeed, this was to become the aircraft's main task in all its theatres for the next few years.

In August 201 Squadron move its twelve Sunderlands from Castle Archdale to Pembroke Dock as part of a general reorganization that saw a number of flying boat squadrons allocated a post-war role of transport flying. The ORB recorded that the unit had, by September, 'reverted entirely to peace-time activities; the next two years were to prove problematical for this and many other units as demobilisation created havoc'.

There were some good deals in the immediate post-war months :

> Two aircraft alighted at Schleswig, the Squadron's first appearance in Germany. Food was excellent, messing charges ½ Mark (3d) a day, and at the first rate Officers Club on camp dinner cost 3 Marks, a bottle of champagne 32 Marks and liqueurs ½ Mark each.

One of the first major 'trooping' tasks saw 201 Squadron carry 800 personnel to Alness in Iceland – and repatriate 1,000 men. Six aircraft were used on this task, although additional sorties had to be mounted by aircraft of 4 OTU. In the same month, October, the squadron almost doubled in strength, acquiring ten Sunderland Vs from 10 Squadron RAAF as this unit disbanded. Repatriation sorties were primarily flown to Iceland, Norway, Gibraltar and Cairo.

August brought the demise of the two Canadian squadrons: 423 Squadron bade farewell to its Sunderlands and moved to Bassingbourn to become a Liberator transport squadron (it then disappeared from the RAF list on 31 August), whilst sister-unit 423 Squadron followed a similar path, flying its last operation on 2 June (two aircraft flying an Adder patrol for a convoy), moving to Bassingbourn to re-equip with Liberators and disbanding on 31 August.

January 1945 and a CA-coded aircraft of 131 OTU is a victim of the 'great freeze' at Killadeas. Ken Delve Collection

In December the flying boats had a final 'shot' at German U-boats, Coastal Command mounting Operation *Deadlight* during which aircraft from a number of squadrons carried out attacks on surplus enemy submarines. Typical of these sorties were those flown by two aircraft of 201 Squadron on 10 December:

> At 1200 H/201 attacked the third and last U-boat allocated for the day. Six 250lb Torpex DC XI were dropped from 500ft in a 275ft stick spaced at 55ft – all undershot and there was no visible damage. Five minutes later B/201 attacked with a perfect straddle causing severe damage to the U-boat, which settled stern down. (201 Squadron ORB, December 1945)

Sunderlands monitored the surrender of German U-boats and escorted the vessels to Allied ports; here a 461 Squadron aircraft escorts a 1,200-ton boat. Ken Delve Collection

The last operational Sunderland sortie of the war by Coastal Command: 3 June convoy patrol by Wg Cdr Barrett of 201 Squadron. Ken Delve Collection

Sunderland V RN271 of No. 4 (C) OTU; by early 1945 the OTU had received Sunderland Vs but had also begun to reduce its training throughput. Andy Thomas Collection

The 'Other' Flying Boat

Only two flying boat types were used by the RAF for offensive operations during the Second World War – the Short Sunderland and the Consolidated Catalina – so it seems worth a quick note about the latter as part of the overall picture of operations.

March 1941 saw the first Catalinas enter RAF service, 240 Squadron at Stranraer being the first to equip, this unit giving up its Supermarine Stranraers. The Catalina had its origins in a mid-1930s US Navy requirement for a long-range patrol-bomber, and the type duly entered service in 1936 as the PBY-1. An advanced design powered by two Pratt & Whitney Twin Wasp R-1830s, the aircraft was an immediate success and it was inevitable that it would attract the interest of the British Air Ministry during their late 1930s' search for modern aircraft with which to re-equip the RAF. Thus, a single aircraft, designated P9360, arrived at Felixstowe in July 1939 for evaluation by the Marine Aircraft Experimental Establishment (MAEE).

An initial order for thirty aircraft, to be designated Catalina I, was placed in late 1939; by the end of the war the RAF's Catalina buy had reached well over 500 aircraft and, in seven variants, these aircraft served with twenty RAF squadrons at home and overseas. Like the Sunderland, the Catalina served in all the maritime patrol and anti-submarine roles and scored a number of successes against U-boats. Indeed, two Catalina pilots were recipients of the Victoria Cross for anti-U-boat attack, an award that eluded the Sunderland pilots despite their many valiant attacks.

Peacetime Routine

The UK establishment gained another Sunderland squadron in April 1946 when 230 Squadron moved to Calshot from Singapore, joining the Coastal Command routine of Meteorological sorties (Epicure), training and exercises. In the same month, 201 Squadron moved to Calshot.

One of the annual requirements for the UK-based Sunderland squadrons was participation in the Anti-Submarine Courses run at Castle Archdale. This unit became the Joint Anti-Submarine School (JASS) and crews attended a 6–8 week course.

By August 201 Squadron had been reduced to six aircraft and six crews again, but even so it was difficult for the Squadron to maintain any level of crew operational efficiency as demobilization had a major effect on aircraft serviceability, as well as the loss of aircrew:

> [In the] tremendous post-war upheaval of personnel, serviceability of aircraft was still a problem. Further attempts were made to stabilize crews and train them to a standard suitable for any tasks the Squadron might be called upon to perform. (201 Squadron ORB, February 1947)

Sunderlands in storage – minus engines and fins – at Sydenham. Peter Green Collection

Ken Robinson was introduced to the Sunderland V at 302 FTU, Oban, and was immediately impressed:

> I thought the new aircraft was superb. When the throttles were opened, I felt as though I was being kicked in the back and the general feeling of power was incredible compared with earlier versions. There were greater revelations to come when Flt Lt Peter Moffat feathered both port engines and the aircraft continued to hold height at over 112kt (Ken Robinson, *ibid*).

L. Still did his national service at 57 MU Wig Bay as an FME (Flight Mechanic Engines):

> I was delighted to find many Sunderlands at Wig Bay, they were drawn up in lines on dry land for storage, most of them never to fly again. Every now and then one would be relegated to the breakers yard and sold to a civilian for £10. The firm would come and make short work of the aircraft with oxy-acetylene burners. But first of all it was my job to de-fuel the aircraft. This was done by winding an emergency handle clockwise above the head of the pilot and co-pilot, the fuel would gush out through a pipe in the hull and soak up in the field below. It had to be left for well over a week before the fumes had cleared and the

Sunderland III ML827 of 330 Squadron at Sullom Voe in April 1945; the Squadron operated Mk IIIs from February 1943 to May 1945. Andy Thomas Collection

Distinctive 'servicing sheds' for the Sunderland providing a covered area over the engines; PP163 at Wig Bay in 1956, its last unit having been 235 OCU. Peter Green Collection

Amongst other units to operate Sunderlands post-war was the ASWDU.
Peter Green Collection

> burners could be lit with safety. We wasted hundreds of gallons of 100 Octane fuel in this way. In fact we used to collect some of this fuel in 50gal oil drums and wash our battle-dresses and overalls in it. It was a good cleaning agent but we had to leave our clothes airing for a week before the fumes cleared and we could wear them again.

The main training aim was that of maintaining proficiency in anti-submarine warfare, squadrons continued to detach crews once a year for the 6–8 week course at the Joint Anti-Submarine School (JASS) at Londonderry. Realistic training profiles provided excellent training value. There were, sadly, still losses; during the June–July 1947 JASS course 201 Squadron's PP113 crashed after completing a simulated attack on the submarine HMS *Sentinel*, Fg Off Clark and crew being killed. February 1948 saw 230 Squadron participate in the fist 'Flying Fish' exercise, co-operating with submarines out of Portland, dropping sonobuoys, tracking and attacking the submarines. The squadron recorded numerous problems with the sonobuoys, the short life of the batteries being put down to their length of time in storage.

Berlin; 1948. Sunderlands provided a small but important part of the Airlift.
Peter Green Collection

Berlin Airlift

July 1948, 230 Squadron had just 'finished the Flag Officer Submarine's war when told to return to Calshot where orders were received to proceed at once to Hamburg for Operation *Plainfare*'. The decision by the Russians to close all land access to the sectors of Berlin controlled by the Western Powers was to lead to an increase in East-West tension – and an impressive feat of air power with the air supply of the city. The Sunderland participation was small-scale in terms of the overall air effort, but was nonetheless significant.

Indeed, 2 July brought a warning order to a number of squadrons for Operation *Plainfare*, the Berlin Airlift. Four Sunderlands of 230 Squadron and one of 201 Squadron moved to Hamburg the same day as the advance party for what would become the major tasking of both squadrons for the rest of the year. The rest of 201 Squadron arrived on 6 July:

> [There were] early difficulties with refuel and reloading as no RAF refuellers were available and aircraft were either moored alongside a tanker or else a duck came alongside with 40gal drums which had to be hand pumped. Eventually RAF refuellers arrived, as did short type buoys – which greatly relieved the captains, who were afraid of damaging their aircraft on the heavy iron buoys originally used. Various loads were carried initially in an attempt to save refuel after every sortie, but later the load was standardized at approx 10,000lb. Cargo consisted of noodles, cigarettes, flour, yeast, spam and tins of M&V. A number of passengers were carried out of Berlin. (201 Squadron ORB, July 1948)

The first Berlin flights were flown on 7 July by 'D' of 201 Squadron (Flt Lt Hackman).

The Sunderlands were based on the River Elbe at Finkenwarder and were to operate into the Havel See in Berlin. Fifth July saw 230 Squadron's Wg Cdr Crosbie fly the first Sunderland (VB887) to the Havel See. Along with 235 OCU the Sunderland detachment was known as Coastal Command Detachment (Finkenwarder) BAFO (British Air Force of Occupation), the detachment using the facilities of 22 Heavy Workshops REME (which in fact were the ex-Blohm und Voss aircraft factory). According to 230 Squadron,

> Water handling was extremely hazardous at first due to pleasure yachts on the Elbe and Havel; later these were prohibited from flying areas,

With Gibraltar as a backdrop, 230 Squadron's 'X' (now wearing the squadron code '4X') poses in a post-war shot. W/C T Holland

> but wrecks and uncharted sandbanks still proved hazardous. Aircraft were moored to large iron marker buoys and mooring up without reversible pitch propellers was an exciting business. The Squadron experimented with various food loads such as tinned meat, macaroni, dehydrated potato, dried eggs and milk, and cigarettes. Standard load was 10,000lb of freight and 700gal of fuel, although this meant that aircraft had to refuel after every sortie – a task that was carried out by hand until sufficient refuellers became available. The cargo was landed by DUKWs and various small craft.

Similar comments were made by 201 Squadron.

Between 6–31 July, 230 Squadron flew 125 sorties, sometimes three a day by a single aircraft. August saw 235 OCU replaced by Hythes of Aquila Airways, their civilian crews being mainly ex-RAF.

From August onwards a number of the outbound sorties carried documents; this almost certainly refers to German wartime material although no details are given. The airlift continued into the autumn and by October much of the cargo was salt, the Sunderlands being the only aircraft types with adequate anti-corrosion protection for this material.

In September 201 Squadron flew eight sorties to Berlin in which they carried 380 tons. Salt was the main commodity carried in October. By November some crews were flying three sorties a day and the turn-round times had been reduced at Havel and Finkenwarder; at Havel the record was twelve minutes from touchdown to take-off whilst at Finkenwarder, which included refuel, it was twenty minutes. On 22 November, 201 Squadron brought out its first passenger load of undernourished children and from that date to the end of the month the squadron carried 683 children and fifty-six nurses. The Sunderlands of 230 Squadron also brought out sick adults and undernourished children. With the end

A 201 Squadron Sunderland V performing at a Battle of Britain display in the early 1950s, note both starboard engines are feathered. Via Arthur Banks

of the detachment on 15 December, 230 Squadron returned to the UK having flown over 1,000 sorties and carried 5,000 short tons of supplies to Berlin.

Tom Holland flew for six months of the airlift with 230 Squadron:

> Finkelwarder was not a very good site for our operations, conditions were quite primitive and the river Elbe was not ideally suited for Sunderland operations – it had a strong tide and was busy with ships ranging from ocean-going transports to small ferries. We had an average of 12–14 aircraft available from the two squadrons and 235 OCU, most of which flew around three sorties a day each. We started work at 0430 in the summer months so as to get the maximum amount of daylight – we were not able to operate at night – and our routes down the corridors were at 1,000ft. The average load was around five tons and our most distinctive load was salt – we were the only aircraft type able to carry this important commodity. The Russians were always an unknown quantity, fighters would buzz us, making mock attacks and flying fairly close – we would raise our coffee cups as if to offer them a drink! Landing at Berlin could be interesting as the lake was quite small – OK if there was a wind but in calm conditions with a full load it was pretty tight, a few Sunderlands ending up clipping the reeds around the lake shores.

Operation *Plainfare* sorties ended in mid-December for the Sunderland units; the combined Coastal Command Sunderland force had flown 1,000+ sorties, eleven

The training task in the 1950s was in the hands of 235 OCU based at Calshot; ML817 spent some time with this unit. Andy Thomas Collection

crews having each notched up over fifty sorties, with 201 Squadron's Flt Lt Cooke recording 110 sorties.

The Coastal Command Sunderland squadrons settled down to a routine of exercises and training, these having been abandoned during the six months of the Berlin Airlift. On 17 January 1949, the advance party of 201 Squadron moved to Pembroke Dock, being joined the following month by 230 Squadron. Meanwhile, in May all five of 201 Squadron's aircraft were at Shorts for modification with sonobuoy receiver equipment, this modification programme continuing into June. However, not everyone was happy, 201 Squadron's ORB of June 1949 noting that 'the modification installation is satisfactory but it is felt that the separation of the sonobuoy operator from the flight deck is undesirable'.

Typical activity for the year included exercises such as *Lancer* and *Verity*, *Porcupine*, *Flying Fish*, participation in Flag Officer Submarine's Summer War, and PR visits to seaside towns.

One of the few notable events to involve Sunderlands in the early 1950s was the 1951 expedition to explore Queen Louise Island, Greenland, aircraft from 201 Squadron providing support from 23 July to 29 August. A second expedition took place the following summer, again with Sunderland support.

By the mid-1950s, Unit Establishments had fallen to five or six aircraft. For 230 Squadron at Pembroke Dock the routine comprised continuation training, Command categorization, naval exercises (such as *Dawn Breeze*), plus the occasional ASR sortie or detachment to Gibraltar. It had already been decided that the flying boat squadrons would disband in February 1957; the squadrons at Pembroke Dock, 201 and 230, flew their remaining aircraft to the Flying Boat Storage Unit at Wig Bay. The disbandment parade at Pembroke Dock on 28 February brought the Sunderland's UK-based role to an end. The Sunderland had established its place in RAF history – in war and peace.

The 1951 Greenland Expedition was supported by 201 Squadron; RN299 at Seal Bay.
Peter Green Collection

Foreign Users

Australia – RAAF

Australia was a major wartime operator of the Sunderland; the role of 10 Squadron RAAF, one of the most operationally effective Sunderland units, has been included in the main account above. Likewise, the wartime operations of 461 Squadron have also been covered above. January 1944 had seen the departure of two Sunderlands from Mount Batten en route down to Australia, and these aircraft became part of the initial equipment of 40 Squadron RAAF, which was formed at Townsville on 31 March. The Squadron operated transport, primarily freight, services for Allied forces in Australia, the main routes being Port Moresby and Darwin. The squadron gave up its Sunderlands in the middle of 1946, five of the six aircraft going to Trans-Oceanic Airways (although most had been scrapped by the early 1950s).

Australia – RAAF

Registration	*RAF Serial*	*Acquired*
A26-1	ex-ML730	Nov 43
A26-2	ex-ML731	Nov 43
A26-3	ex-ML732	Nov 43
A26-4	ex-ML733	Nov 43
A26-5	ex-ML734	Nov 43
A26-6	ex-DP192	date unknown

(the serials include only those aircraft with Australian registrations)

Canada – RCAF

Two Royal Canadian Air Force (RCAF) squadrons – 422 and 423– operated Sunderlands from the UK late 1942 to the end of the war. The operations flown by these two units have been included in the main account above.

France – Aeronavale

A number of French crews flew Sunderlands during World War Two; a Free French unit, 4th GR Squadron, received Sunderlands in February 1943 (later becoming 343 Squadron), details of which are included in the West Africa chapter.

Post-war, nineteen aircraft were reconditioned at Belfast by Shorts and delivered to the Aeronavale in 1951, for service with Escadrille 7FE at Dakar. The last of these aircraft was retired in 1960, one of which, ML824, was gifted by the French Government to the Sunderland Trust for preservation at Pembroke Dock, arriving there on 24 March 1961.

New Zealand – RNZAF

The first Sunderland use by a New Zealand unit was by the previously Catalina-equipped 5 and 6 Squadrons, both of whom operated a small number of Sunderland IIIs from late 1944 (only four aircraft in total). Post-war, from November 1947, these four aircraft were operated by New Zealand National Airways Corporation for four years before returning to the RNZAF and final scrapping in 1954.

The RAF had acquired an impressive array of 'retired' Sunderlands at Wig Bay where their fate was usually the scrapman's

France – Aeronavale

Registration	*Acquired*
ex-ML739	Feb 52
ex-ML750	Jul 47
ex-ML757	Jun 51
ex-ML764	Jul 51
ex-ML778	May 51
ex-ML779	Apr 51
ex-ML781	Apr 51
ex-ML796	Aug 51
ex-ML799	Jun 51
ex-ML800	Jul 57
ex-ML816	Aug 51
ex-ML819	Sep 51
ex-ML820	Nov 51
ex-ML821	Sep 51
ex-ML824	Oct 51
ex-ML835	Jun 45
ex-ML841	Jun 45
ex-ML851	Jun 45
ex-ML854	Jun 45
ex-ML866	Nov 51
ex-ML872	Dec 51
ex-ML874	Jun 45
ex-ML877	Jan 52
ex-RN284	Dec 57
ex-SZ571	Nov 57
ex-SZ576	Jul 57

The French Aeronavale became a major post-war user of the Sunderland, the main unit being Escadrille 7FE at Dakar. Peter Green Collection

The RNZAF operated a number of Sunderlands post-war; NZ4116 in service with 5 Squadron.
Peter Green Collection

axe; however, sixteen aircraft were refurbished here and supplied as Mk Vs to the Royal New Zealand Air Force in 1953. These aircraft equipped 5 (MR) Squadron based at Hobsonville (Auckland) and Lauthala Bay (Fiji). The other maritime unit, 6 Squadron at Hobsonville (Auckland), also acquired a number of aircraft, as did the Maritime OCU. The last of these were withdrawn from service in spring 1967 and had the distinction of being the last operational Sunderlands. One aircraft, albeit not one of these last five, is displayed at the Museum of Transport and Technology, Western Springs, Auckland.

New Zealand – RNZAF

Registration	*RAF Serial*	*Acquired*
NZ4101	ex-ML792	Oct 44
NZ4102	ex-ML793	Oct 44
NZ4103	ex-ML794	Oct 44
NZ4104	ex-ML795	Oct 44
NZ4105	ex-PP110	Jan 45
NZ4106	ex-RN280	Dec 53
NZ4107	ex-VB883	Nov 53
NZ4108	ex-ML814	May 53
NZ4109	ex-DP191	?
NZ4110	ex-PP129	Jul 53
NZ4111	ex-VB880	Jul 53
NZ4112	ex-VB881	?
NZ4113	ex-PP124	Oct 53
NZ4114	ex-SZ561	Oct 53
NZ4115	ex-SZ584	Sep 53
NZ4116	?	?
NZ4117	RN286	Oct 53
NZ4118	RN305	Oct 45
NZ4119	PP143	Jan 45
NZ4120	RN291	Feb 54

Norway

During the wartime period the Norwegian squadrons adopted numbers within the RAF sequence; the operations of 330 Squadron have been included in the main Home Water narrative.

Portugal

The Portuguese Navy acquired Sunderland P9623 when this aircraft force-landed in Portugal on 14 February 1941.

South Africa – SAAF

Sunderland Vs had been delivered to 35 Squadron at Congella, Durban in spring 1945.

The immediate post-war task for the SAAF Sunderlands was trooping – in this case from the Middle East back to South Africa. Most of the Sunderlands had been sold for scrap in March 1955 but three aircraft survived a few more years and the last SAAF Sunderland flight took place on 8 October 1957 (aircraft 1710).

South Africa – SAAF

Registration	*Acquired*
ex-PP104	May 45
ex-PP109	Jun 45
ex-PP125	Sep 45
ex-PP152	Apr 45
ex-PP156	May 45
ex-RN279	Jun 45
ex-RN281	Jun 45
ex-RN295	Jul 45
ex-RN296	Jul 45
ex-RN305	Oct 45
1709 ex-ML798	May 45

CHAPTER FIVE

Middle East and Mediterranean

The impressive bay at Gibraltar was home to Sunderland detachments, and 202 Squadron, for much of the war. Ken Delve Collection

The Mediterranean could, from the anti-submarine war perspective, be divided into east and west, the west end having a direct connection with the submarine war in the Atlantic whereas the east end (from Malta to Egypt) was more concerned with operations in Greece, Italy and North Africa. However, for our purposes the Mediterranean is treated as one theatre of operations. Prior to the entry of Italy into the war in June 1940, the area of most concern was the western Mediterranean, with Gibraltar as a significant airfield. The roles of 'the Rock' and the island of Malta were critical to Allied strategy (and survival) in this theatre. Maritime units based at Gibraltar were usually retained under the operational control of HQ Coastal Command, a slightly unusual procedure but one that reflected the part these units played in the Atlantic War.

On 26 April 1939, No. 228 Squadron at Pembroke Dock was warned that its aircraft would be handed to 202 Squadron, then operating out of Gibraltar with Catalinas, whilst it would subsequently re-equip with another type.

Two aircraft left Pembroke Dock for Malta on 29 April, although L5807 had to force-land at Berre with engine failure. The original plan was soon modified as on 3 May a warning order was received from No. 16 Group that 228 Squadron would deploy to Alexandria, as part of No. 86 Wing, to augment RAF Near East. With the ground party boarding HMT *Dumana* on 9 May, the move went ahead, the intention being that this ship, along with the refueller SS *Pass of Balmaha*, would be stationed at Alexandria as the squadron's base. During the early part of the summer the squadron visited a number of potential areas to check their suitability for either servicing or operations: Fg Off Burnett took N6133 to the Sea of Galilee in July to ascertain the possibility of compass swinging by manhandling the aircraft in the shallow water, whilst N9070 flew to Lake Tiberias, this being one of the freshwater locations used for washing down the Sunderlands. Later in the month, Wg Cdr Barnes flew N9020 on a survey of the Nile to look at possible compass swinging and wash-down areas at Lake Quaran; however, the starboard engine caught fire and the aircraft did not return until early August after an engine change had been carried out:

> Under difficult conditions, there being no facilities for engine changes, and the terrific heat making it impossible for the airmen to work during the day, the result being that most of the work was done by night. (228 Squadron ORB)

The outbreak of war saw five aircraft at Alexandria and three at Malta – all serviceable and fully bombed up. Having spent a year becoming familiar with the needs of the Mediterranean theatre, the squadron had mixed feelings when ordered back to the UK, four more aircraft moving to Malta on 9 September. The following day four Sunderlands flew back to Pembroke Dock via Marignane.

Meanwhile, 230 Squadron had arrived at Alexandria in early May 1940 from the Far East, the CO, Wg Cdr Bryer, flying to Cairo on 13 May for a conference with HQ Middle East and No. 201 Group (which Bryer, on promotion to Gp Capt, took over in June), at which it was confirmed that the squadron's main task would be the 'anti-submarine protection of the Allied Mediterranean Fleet whilst at sea'. Mersa Matruh lagoon was one of a number of locations examined for use as advanced landing bases

Italy Enters the War

On 10 June Italy declared war on Great Britain and operations commenced in the Mediterranean and Western Desert against this new enemy. Italian naval power in the Mediterranean was far greater than that of Britain, and included over 100 submarines; furthermore, much of the Italian equipment was modern and was thus a significant threat.

By 12 June two serviceable aircraft were based at Alexandria, using HMT *Dumana* as a base ship, with offices and stores at the nearby Imperial Airways depot. Flt Lt Ware (L5806) flew the first operational sortie, an ASP, the following day. having flown a variety of operations, including reconnaissance of Tobruk and searches for the Italian fleet. The first U-boat attack was carried out on 28 June, Wg Cdr Nicholetts (L5806) dropping three bombs on a large submarine, the bombs overshooting the target, although it appears from subsequent revelations that his attack damaged the *Anfitrite*, causing the submarine to return to base for repairs. The following day he attacked a small submarine, dropping four bombs in his first attack, all of which failed to explode; two more bombs were dropped in a second attack, these falling 20 yards ahead of the submarine track but with no apparent effect.

Meanwhile, the first operational sorties were flown on 10 June, two aircraft carrying out ASPs ahead of an eight-destroyer

L2159 rides at its mooring in Marsaxlokk Bay, Malta, as 230 Squadron passes through on the way to the Far East. Ken Delve Collection

force sweeping for submarines to the west of Alexandria. The patrol operated from dawn to shortly before dusk, the first pair of aircraft being relieved by a second pair at midday. This task continued the following day, with other aircraft tasked on reconnaissance sorties looking for minefields. If an air raid warning was received at Alexandria, the Sunderlands were 'dispersed' by being taxied around the Bay, with all guns manned! On 20 June, Sunderland L2160 of 230 Squadron was tasked to recce Tobruk and during the sortie was attacked by four CR.42s; during a 15-minute engagement one Italian fighter was shot down but the Sunderland received numerous hits from explosive bullets, including a large hole in the hull and holes in the fuel tanks.

Aircraft were deployed to Malta as required to give added protection to eastbound convoys. The first success for 230 Squadron came on 28 June, Flt Lt Campbell, airborne from Aboukir in L5804, finding and sinking the *Argonauta*. The following day the same crew were successful again, the aircraft was on a reconnaissance to the west of Zante when it picked up the submarine *Rubino*; she was sunk in the attack and the Sunderland picked up four survivors. An Italian report stated that two bombs had hit the stern and conning tower causing the boat to sink very rapidly. The Sunderland, out of bombs, machine-gunned another submarine found on the return journey. The following day this same pilot was involved in a dive-bomb attack on a destroyer in Augusta harbour, the elevator fabric of L5803 being damaged by the high-speed dive! Another bombing attack was made on 1 July when L5803 was en route to Egypt from Malta, the target being a destroyer near Tobruk, the four bombs missed but the aggressive intent of 230 Squadron and Flt Lt Campbell was certainly evident. Indeed, the aggressive activities of British naval and air forces was in large measure responsible for the poor showing of the Italian fleet, its commanders showing a marked reluctance to risk engagement.

At the end of June 228 Squadron detached two aircraft to join 202 Squadron at Gibraltar, one of their main tasks being reconnaissance of French naval bases such as Oran, Mers-el-Kebir and Algiers. The first reconnaissance of these bases was made on 1 July, two sorties confirming the presence of major French warships. On 4 July Flt Lt Brooks in P9621 was on a recce of Oran and Algiers when attacked by three French Curtis 75As; the Sunderland claimed to have shot down one fighter and damaged another, although in return it had been badly damaged and one crew member was injured. The aircraft had to return to Pembroke Dock for repairs. British warships, having failed to negotiate the surrender of the French warships at Oran, opened fire and destroyed or seriously damage most of the major vessels, with heavy loss of life amongst the French sailors. A Sunderland reconnaissance sortie on 5 July confirmed the result of the bombardment.

The squadron also deployed aircraft to Malta; on 9 July Flt Lt McKinley (L5807)

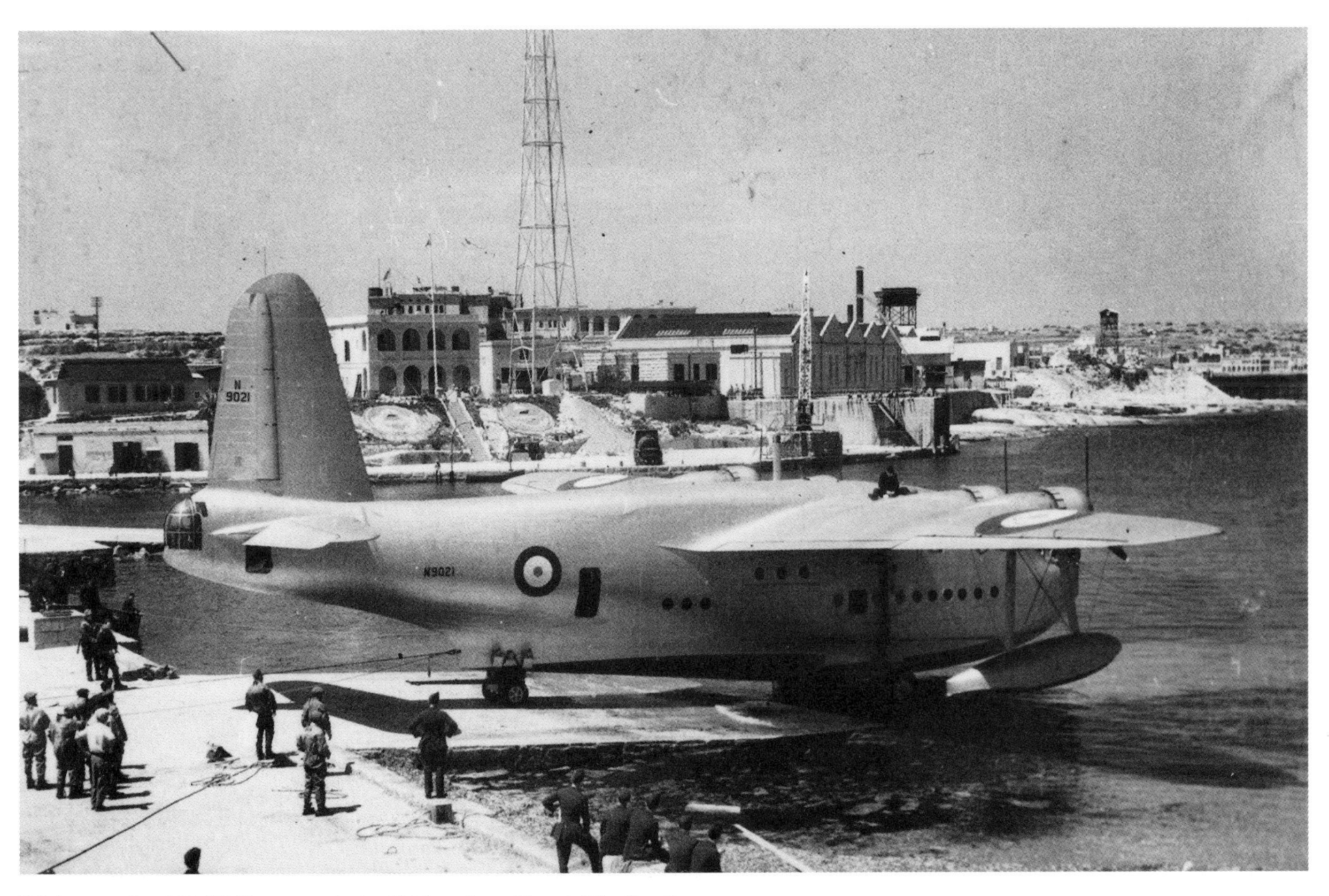

Kalafrana, spring 1939; N9021 was attached to 202 Squadron, although this allocation does not appear on the aircraft's record card. Andy Thomas Collection

took off from Malta on a search for the Italian fleet – and eventually located approximately forty ships which he then shadowed for nine hours. A second 228 Squadron Sunderland (N9020, Sqn Ldr Menzies) took over and flew a further 9-hour shadow, during which time the crew had an inconclusive engagement with an He 115. Three days later Sqn Ldr Menzies was flying a Malta patrol when he came across a U-boat:

> In the first stick three 250lb A/S bombs were dropped and fell close to the stern. Two bombs were dropped in the second attack whilst the U-boat was submerging and these fell abaft the conning tower. A single bomb was then dropped ahead of the submarine. Excessive quantities of air were observed to come up a short distance from the last observed position and this was taken as a final indication of the U-boat's end. (228 Squadron ORB, July 1940)

It is likely that this was the 954-ton *Settimo*, this submarine reporting an air attack on July 13 in which it suffered light damage.

Most of July's effort was spent keeping tabs on the Italian fleet, both squadrons flying similar patrols from Egypt and Malta. 230 Squadron's Flt Lt Woodward claimed to have attacked and sunk an Italian submarine on 7 July, but there appears to be no confirmation of this. In the latter part of the month this pilot and Gp Capt Bryer flew to Port Said, Lake Timsah and Suez to evaluate these locations as potential flying boat bases.

Italian fighters were met from time to time, Flt Lt Garside encountering three Maachi 200s on 28 July, the crew claiming one shot down and one damaged in a 15-minute combat. The same day, another 230 Squadron crew (Sqn Ldr Ryley, L5804) tangled with four Maachis, shooting down one in a 57-minute combat. The Sunderland was badly damaged and three gunners wounded:

> Much credit went to LAC Campbell, a fitter, who remained in the wing of the aircraft, plugging holes in the tanks, until rendered unconscious by petrol fumes. The aircraft landed at 1215 and was beached in a sinking condition. (230 Squadron ORB)

A number of such combats had been reported by Sunderlands of both squadrons during reconnaissances of Italian ports.

The 228 Squadron detachment was tasked in August 1940 with patrols around Crete. A standing task was shadowing of Italian shipping, and on 6 August, N9020 began shadowing an Italian convoy, also taking the opportunity to bomb a tanker, but without result. The shadow task was taken over by Flt Lt Smith in N9025, but the aircraft was shot down, one crew member (LAC Jones) being killed. Having force-landed on the water, the Sunderland maintained W/T contact with its base until an Italian destroyer came alongside and took the aircraft in tow. Meanwhile, the squadron commander had become airborne to effect a rescue, this sortie being aborted when it was learned that the Sunderland crew had been taken prisoner.

Flight Lieutenant Alington (L2166, 230 Squadron) was on ASP on 30 September, being informed by a destroyer that it had attacked and damaged a U-boat. The Sunderland dropped bombs on the area of air bubble disturbance, causing additional damage to the *Gondar*: the submarine was subsequently scuttled by its crew.

On 12 October, Flt Lt McCall (L2164), having been diverted from his patrol to search for a Fairey Fulmar downed by a Cant, located the sinking aircraft and landed to pick up three survivors. The Fleet Air Arm crew was given a hot meal and a change of clothes, the Sunderland completed its patrol and then landed at Kalafrana Bay. Two weeks later, Flt Lt Ware (L5806) was attacked by two Italian fighters, two of his crew being wounded:

> Mattresses and clothing in the aircraft were set on fire, burning articles were thrown out of the rear door. Flame floats and practice bombs were set off by explosive bullets and caused the aircraft to fill with smoke. The rear turret was put partially our of action by having the starboard control handle shot away. The Sunderland was badly holed below the waterline and was taken up the slip immediately on return to Kalafrana. (228 Squadron ORB)

The Italians invaded Greece on 28 October but the operation went badly from the start and ended with an Italian withdrawal in mid-November.

On 2 December, Operation X was put into effect, the intention being to intercept fast Italian convoys running from Italy to Benghazi. The role of 228 Squadron's Sunderlands was to locate these convoys

> ... with the aid of moonlight and ASV. On the captain's discretion the convoy could then be attacked, but a special force of Fleet Air Arm torpedo bombers was held at readiness for such attacks. The Sunderland carried four SAP and four A/S bombs to drop in sticks of four – A/S, SAP, SAP, A/S. (228 Squadron ORB, December 1940)

On 12 December, No. 230 Squadron sent an advanced party to Scaramanga (15 miles west of Athens on Lake Eleusis) as this provided an ideal location from which to mount searches for Italian naval vessels:

> This was an almost ideal flying boat alighting area. The site of the moorings is off a Greek naval base at the east end of the lake. There is a slipway capable of taking one Sunderland; though designed for launching ships the Greek Navy had installed an electric winch for hauling out. The new centrally-heated Drawing Office was used to accommodate officers and men.

The detachment flew patrols around Crete, using Suda Bay as an advanced landing base, the main roles of the Squadron being:

- Searches west of Crete to give security to convoys moving in or through the Aegean Sea.
- Searches in north part of East Ionian Sea when units of Mediterranean Fleet are at sea to give those units warning against approach of enemy ships and generally to give C-in-C Mediterranean an opportunity to come to grips with the enemy.

> During the Greek campaign we had a refuelling ship, the *Pass of Balmaha*, a small tanker. As far as I can remember, this was at Suda Bay most of the time [according to the official records, the ship moved from location to location depending on the task]. We could refuel by taxiing up to her stern and making fast to her bollards, but often in the early days refuelling had to be done manually by hoisting up 4gal drums on to the wing of the aircraft. As our capacity was something like 2,350gal, the time taken to perform this task was pretty lengthy. (Dundas Bednal)

Meanwhile, for 228 Squadron operating out of Malta, the weather in January 1941 was causing trouble. The danger from big swells was such that a skeleton crew of one pilot, one rigger and one wireless operator slept on each aircraft. On 21 January, two aircraft were moved to the more sheltered waters of St Paul's Bay. By mid-January Suda Bay was closed as an advanced operational base and was only used as an emergency landing base; meanwhile, 230

Squadron was flying most of its sorties from Scaramanga. Operations for the latter part of January are detailed in the table below.

Operations against Italian merchant ships and warships remained a top priority into 1941, C3s and Modified C3s being the main types of patrol. In its record for 26 January the squadron noted:

> Modified C3 patrol [Flt Lt Glover, L5807], sighted one small MV and seven self-propelled barges just east of Kerkenah. Reported to base and shadowed for two hours. No striking force sent out and Sunderland was ordered to continue patrol, but nothing sighted. Until the position has been clarified by the Admiralty it was laid down that aircraft may not attack either unescorted merchant ships or unescorted merchant convoys as this infringes International Law. Any ship within 30 miles of any Italian territory in the Mediterranean may however be sunk on sight. During this patrol one Cant 501 was observed to be shadowing the Sunderland but when chased by the latter it made off towards Tripoli first jettisoning two or three bombs.

The following day the Modified C3 patrol flown by Fg Off Lamond (T9048) was more productive:

> Early in the patrol one Italian MV sighted and reported to base, ordered to continue patrol. Sighted aircraft, identified as Ju 52. Sighted one EV and two MV north-east of Kerkenah Bank on southerly course. Reported them to base and shadowed for three hours. ASF of seven Swordfish, escorted by two Fulmars, then arrived and carried out attack. One MV sunk and one hit. Followed ASF back to base. One Swordfish having petrol trouble asked for Sunderland to escort but Sunderland also had engine trouble and had to return to St Paul's Bay. Message from Admiralty: The combined RAF and Naval Aircraft Operations which resulted in successful attack on the convoy at a distance of 160 miles from Malta was well planned and executed. Those concerned are to be congratulated. This provides an excellent illustration of the correct employment of air search and striking forces.

On 4 February 228 Squadron noted that:

> The situation with regard to bombing of unescorted MVs or convoys has been clarified by the Admiralty. Sunderland aircraft may now violate Tunisian territorial waters south of Latitude 3346N to attack any vessel identified as enemy.

Axis air attacks on Malta caused problems for the RAF and the Sunderlands were not

Date	*Aircraft*	*Captain*	*Time*	*Details*
13 Jan	L2166	Sqn Ldr Alington	0525–1455	Ionian Sea
14 Jan	T9050	Flt Lt Campbell	0720–1510	Ionian Sea
15 Jan	N9029	Flt Lt Lynwood	0555–0945	Ionian Sea
16 Jan	L2166	Sqn Ldr Alington	0540–1155	Ionian Sea
17 Jan	T9050	Flt Lt Campbell	0445–1400	Ionian Sea
	L2161	Flt Lt Smith	0534–1419	Ionian Sea
	L2166	Sqn Ldr Alington	0534–1419	Ionian Sea
18 Jan	T9050	Flt Lt Campbell	0520–1200	Ionian Sea
	L2166	Flt Lt Woodward	1535–?	Aegean Sea
19 Jan	L2166	Sqn Ldr Alington	1405–1600	Gulf of Athens
19/20 Jan	L2166	Sqn Ldr Alington	2300–0515	ASP
20 Jan	T9050	Flt Lt Campbell	0515–1750	Ionian Sea
21 Jan	L2160	Flt Lt Woodward	0530–1440	Ionian Sea
22 Jan	L2161	Flt Lt Smith	0520–1430	Ionian Sea
23 Jan	L2160	Flt Lt Woodward	0720–1650	Ionian Sea
	L2166	Sqn Ldr Alington	0710–1620	Ionian Sea
	L5804	Flt Lt McCall	0510–0927	Escort
24 Jan	L2161	Flt Lt Smith	0520–1445	Ionian Sea
25 Jan	N9029	Flt Lt Lywood	0525–1325	Patrol
26 Jan	L2160	Flt Lt Smith	0805–1440	Patrol
26/27 Jan	L2166	Sqn Ldr Alington	1950–0730	ASV search
27 Jan	N9029	Flt Lt Lywood	0950–1500	ASV search
28 Jan	T9050	Flt Lt Campbell	0755–1345	WAA patrol
29 Jan	L2160	Flt Lt Woodward	0755–1310	WAA patrol
30 Jan	L2166	Sqn Ldr Alington	0800–1325	WAA patrol
31 Jan	N2029	Flt Lt Lywood	0950–1230	WAA patrol
	L2160	Flt Lt Woodward	1250–1735	Patrol

No. 230 Squadron operated from Scaramanga (Lake Eleusis) during 1941 – L5804 sank at it moorings on 25 February 1941. Peter Green Collection

immune from such attacks. On 7 March L2164 of 228 Squadron was attacked at its Kalafrana moorings by two Bf 109s; Sgt Jones, acting as boatguard, managed to get his gun into action before being fatally hit. The Sunderland was badly damaged and was in trouble again three days later when two of the squadron's aircraft were attacked by Bf 109s:

> T9046 was damaged and L2164 caught fire ... a party boarded this aircraft and fought the fire which was apparently got under control, but after an interval blazed up again. The machine was taxied inshore and beached but has to be abandoned, and ultimately sank. T9046 was flown from St Paul's Bay to Kalafrana and taken up the slip for inspection.

The 230 Squadron operations continued to support Allied moves around Greece, including, from 3 March, providing cover for Operation MX3 (*Lustre*), the move of Allied forces from Egypt to Greece. One of 230 Squadron's Sunderlands (Sqn Ldr Alington, L2166) assisted a destroyer and Greek Ansons in an attack on a U-boat on 13 March, but without result. This was the forerunner for one of the biggest naval engagements in the Mediterranean theatre, the Italian Battle Fleet sailing on 26 March to intercept Allied convoys. Over the next few days Sunderlands of 230 Squadron shadowed the enemy fleet, passing reports to Allied planners that eventually led to Admiral Cunningham's successful engagement at the Battle of Cape Matapan.

Alan Lywood and Dundas Bednal were airborne in 230 Squadron's 'V' on 28 March 1941 for a reconnaissance south-west of Crete when they came across a number of Italian warships:

> I spotted three dim shapes through the morning mist on our starboard side. I was able to identify the *Pola*, *Fiume* and *Zara*, along with two Cavour-class battleships. We sent a sighting report of three cruisers and two battleships and then started a very long stint of shadowing. ... Alan then said that the Italians were signalling us and, indeed, there were flashes from the ships. However, we suddenly realized that the flashes were far from communicative as they were the flashes of guns firing at us. Shells landed just aft of us in the sea and some burst with ugly black smoke and dull 'crumps' about 100–200yd astern in the air; their range was very accurate. We were flying at only 70ft above the sea. Here was a classic maritime task – to find, shadow and report an enemy force so that our fleet could engage them. The day went on and as dusk approached Alan asked Alexandria if we could land on the sea and shadow from there as our fuel was getting low. This was turned down – quite rightly as we would have been a sitting target. (Dundas Bednal)

March had also been significant for the move of 228 Squadron from Mediterranean Command to Middle East Command:

> A small maintenance party remaining at Kalafrana to complete work on Sunderland L5807 and to service Sunderland aircraft operating from Malta subsequent to the departure of the main party. An advance party, together with certain stores, were moved from Malta by air, the remaining personnel embarking at Valetta on HMS *Bonaventure*, *Calcutta* and *Greyhound* en route to Alexandria. On arrival at Alexandria the aircraft and crews and members of the advance party were accommodated at No. 201 Group but owing to the limited space available, personnel, other than aircrews, were transferred to Aboukir. There was little available accommodation at Aboukir and the Squadron offices were situated in a Beach Hut with workshops in Blenheim cases and outbuildings – the Squadron NCOs had to find a mess of their own. Later, two additional houses were taken over and the NCOs transferred their mess to these while the officers moved to Richmond House. (228 Squadron ORB)

April brought German military involvement in the ill-fated Italian campaign in Greece, the opening move being the air bombardment of Yugoslavia on 6 April. Ten days later, as Yugoslavia collapsed, two Sunderlands of 230 Squadron flew to Kotor to evacuate Yugoslav leaders, including King Peter. The German forces then swept into Greece. By 12 April German minelaying had made Scaramanga too dangerous for the Sunderlands and so the detachment moved to Suda Bay, although the Lake Eleusis site remained in use for some operations. A number of special sorties were also flown, such as those of 22 and 23 April to evacuate the Greek royal family. Sunderland 'P' was destroyed at its moorings on the lake when bombed and set on fire by Ju 87s on 23 April, the crew escaped unharmed and the midships gunners put in a claim for one Ju 87. The general evacuation of British personnel kept the squadron busy for the next few days, the Sunderlands also acting as escort for BOAC aircraft engaged on the same task. Typical of the late April operations were the following by 228 Squadron:

L5807 was strafed by Bf 109s at Kalafrana on 27 April 1941. Peter Green Collection

> April 24, L9046, Flt Lt Frame was ordered to fly to Nauplia Bay in the Bay of Argos to evacuate RAF personnel believed to be in the district. On arrival just before dusk it was found that the RAF party had moved on. However, after considerable delay twenty-five passengers were taken on board including General MacCabe and one stretcher case. The Captain decided to wait until daybreak before taking off. At dawn

Flt Lt Frame found that the whole Bay was enveloped in dense black smoke caused by an ammunition ship and a troopship which had been bombed in the harbour the previous afternoon. After taxying around for a considerable time to find a clear path, the Captain decided to make a blind take-off on a course given to him by his navigator, Fg Off Austin. The take-off was accomplished without mishap and the evacuees were landed at Suda Bay.

April 25 L9048, Flt Lt Lamond, was ordered to search for a party of RAF personnel believed to be in the Githeon area. A landing was made in the harbour of Githeon as people on the quayside were believed to be the party in question. After anchoring, Greek officials arrived in rowing boats and explained in French that the RAF had moved on to a bay a little further south-west; one Greek Flight Lieutenant was taken aboard and the search resumed. On reaching the position indicated, the aircraft was attracted by flashes from a hand mirror used by the ground party. After flying low over the position on two occasions it was confirmed that the people below were those the aircraft was seeking. A landing was made and the aircraft anchored about 100yd from shore. Contact was made with the party, which comprised 101 officers and men of 112 Squadron. Fifty-two of these were evacuated; the Captain left a message with the Officer Commanding that he would return to pick up the others if he received no other instructions. If he did not return by 8 o'clock the remainder were to get away in the caique which was at their disposal.

April 25, T9046, Flt Lt Frame was sent to Kalamata area where the main body of RAF personnel under G/C Lee were supposed to have collected. A search was made of the area during which the aircraft was fired on from the ground – but on being recognized firing ceased and RAF members appeared from their hiding places to attract the attention of the aircraft's crew. Though the crew had not seen the party, a landing was made in the harbour and enquiries made; during this delay the Group Captain and fifty members of the party had come down from the hills. This party was boarded and flown to Suda Bay.

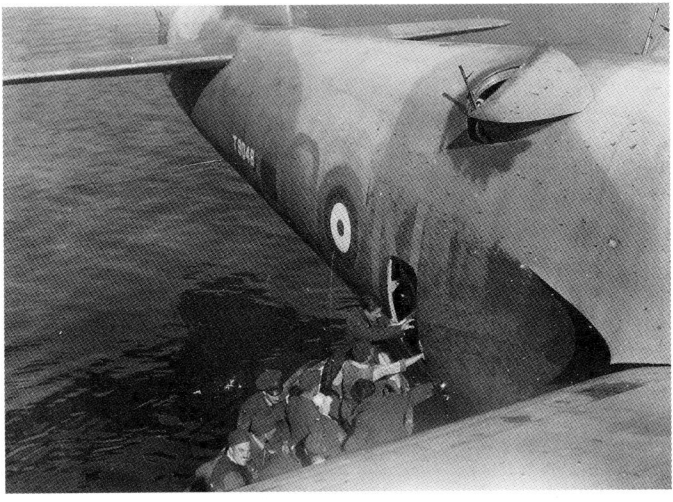

Series of shots of 230 Squadron aircraft during the evacuation of Allied personnel from Crete. Via Arthur Banks

April 25: reports had been received from T9046 that a large party of RAF personnel were in the Kalamata area and Flt Lt Lamond, T9048, was ordered to this area instead of returning to Githeon. At Kalamata seventy-two members of the party were taken aboard, this is a record number of passengers for a Sunderland. This large number was arranged in such a way that the stability of the aircraft was not affected. No bombs were carried and only 350–400gal of petrol was aboard. The take-off was not excessively long, even shorter than the usual run with a full military load.

April 25: On returning to Suda Bay T9048 was at first ordered to refuel to 1,800gal prior to taking off the next morning on a patrol. However, a message was flashed from the shore to refuel to 700gal only and to return that night to Kalamata to deliver a message from the AOC RAF Greece to G/C Lee. The take-off was made without flare path. At approximately 2330 the aircraft was over the harbour at Kalamata; as there was no flare path available the landing was made by landing light. The condition of the sea was calm. On landing, the aircraft crashed and broke up immediately. A portion of the hull remained afloat and in this the four survivors remained until calls for help attracted a small Greek fishing boat.

Three of the survivors were sent to hospital – and were subsequently taken prisoner.

The evacuation continued with our loads getting larger and larger. There was no time for finesse in loading or any observance of maximum all-up weight limits. On one particular journey from Suda Bay to Alexandria 'Y' was loaded up with, as far as I can recall, sixty-three men and 4,600lb of Hurricane ammunition which, I believe, is a record for a Sunderland I. The sea was just below the bottom of the forward hatch when I decided to attempt take-off. I taxied to the entrance of Suda Bay and turned into wind. The sea had a nice ruffled surface and the wind was about 15kt. The first attempt failed and for the second attempt I taxied beyond the entrance to the Bay, this would give us about 5 miles to get off. This time 'Y' staggered into the air with just enough height to safely clear the hills west of the Bay. (Dundas Bednal)

Malta, too, remained a very dangerous place. On 27 April 1941, Plt Off Rees of 228 Squadron was tasked to rendezvous with and escort eight Hurricanes and a Fulmar to Malta, the aircraft having flown off a carrier:

The aircraft arrived 30 minutes late. The arrival at Malta coincided with an air raid and the Captain decided to make a landing as quickly as possible with the intention of getting the boat into the hangar. However, this proved unsuccessful, for after mooring up in the Camber and adjusting one of the beaching legs, the mooring and maintenance party were compelled to leave the boat owing to enemy aircraft being overhead. Whilst the party were making for the shore two Me 109s dived out of the sun above Kalamata, attacking from a westerly direction. Before any protection could be given they strafed the boat and caused it to burst into flames immediately. The boat was towed in a burning condition out of the Camber, but sank near the entrance when the starboard wing fell off. (228 Squadron ORB, April 1941)

The following day the ORB records that Wg Cdr Nicholetts, along with four ASV technicians, was airborne in L5806 to test out the new ASV II installation.

10 May saw Wg Cdr Nicholetts conducting a test on the new twin guns fitted in the midship turret 228 Squadron Sunderland L5806: 'Fg Off Mason stated that the cone of fire was good but the balance of the guns was bad making their movement extremely difficult. Further experiments are being carried out to remedy this fault.' Sadly, but typically, the squadron ORB gives no further details of either the modification itself or the results of the further experiments. Later in the month the squadron conducted the first airborne compass swinging trials (on T9046) using the new astro compass. Courses were checked every 20 degrees, making a total of eighteen courses flown with each course lasting four minutes to give the navigator, Fg Off Austin, ample time to change readings on the astro compass. Again, there are no additional details provided in the official record.

Following the evacuation of Greece, Allied troops had been moved to Crete in some numbers, and on 20 May the German airborne assault on the island began. In a dramatic and ultimately successful (though costly) assault the Germans soon took the island. The Allies were able to evacuate significant numbers of personnel, the Sunderlands once again taking a share of this task.

On 9 June one of 230 Squadron's Sunderlands flew to Erempoli to investigate reports that 1,000 British troops were still at large in the vicinity. The aircraft landed east of Paxmada Island and taxied in to the area; it was hit by gunfire and so taxied clear and took-off. This was the squadron's final participation in the evacuation.

In a co-operative sortie with the Royal Navy on 15 June, one of the squadron's aircraft located and shadowed the French destroyer *Caevalier*, passing reports until the attack force of 815 Squadron Swordfish arrived to sink the vessel. Four days later, 230 Squadron completed its move to Aboukir where it took over the buildings vacated by 228 Squadron and came under the control of No. 201 Group. The move had been made as the 'risk of collision in the over-crowded Alexandria harbour had become too great'. Convoy escorts and ASPs kept the squadron busy throughout the eastern Mediterranean, a number of submarine contacts being made. On 1 August Flt Lt Brand (L2166) was shot down when attacking an Italian submarine in the Gulf of Sollum, the survivors being taken on board by the Italians. The rest of the year was, however, fairly quiet, although Sqn Ldr Kenneth Garside recorded two attacks, one on 22 October and one on 11 November, neither of which produced any result. The year ended with another loss; on 22 December, Flt Lt Hughes (T9071) was en route to Malta with freight and passengers when the aircraft was attacked by two enemy fighters, the Sunderland having to crash-land three miles from Ras al Amar.

The changing nature of the strategic scene saw 228 Squadron depart the Mediterranean Theatre in June en route to West Africa, leaving 230 Squadron at Aboukir as the sole Sunderland unit in the Eastern Mediterranean. However, other units, 10 Squadron RAAF in particular, continued to mount detachments at Gibraltar.

On 1 August, Flt Lt Brand (L2166) made an unsuccessful attack on the submarine *Delfino*, which managed to down the Sunderland. The survivors were picked up by the submarine. In the same month the Germans decided to deploy significant number of boats to the Mediterranean, although these tended to operate more to the west end of the area.

Sqn Ldr Garside scored an early success in 1942, sinking U-577 on 9 January; in the space of a few weeks this experienced Sunderland captain made several more attacks on U-boats. The 230 Squadron record book for January and February 1942 shows little operational activity in terms of U-boat sightings and attacks, other than the exceptional activities of Sqn Ldr Garside and his crew. Between 9 January and 7 February this officer is recorded as having made attacks on six submarines, although without any definite results, all whilst flying Sunderland W3987:

Jan 9 – Dropped three DCs in first attack and threw out flame float, dropped four 250lb AS bombs in second attack. Huge bubble appeared and large patch of oil with bubbles continuing to come up; the 'submarine was undoubtedly destroyed'.

Jan 27 – In first attack dropped a single AS bomb, in second attack dropped three DCs set at 50ft followed by a third attack with three DCs set at 100ft.

Jan 29 – In a dive attack dropped eight AS bombs in a stick from 600ft, straddled conning tower. After the spray had subsided the submarine had gone, a second or two later the stern rose vertically out of the water and slid under. An oil patch appeared and spread 400 × 200yd.

Feb 6 – Eight AS bombs were dropped in a straddle over the centre of the submarine, after the spray subsided the submarine was seen to have broken up and the disconnected sections were momentarily seen above the surface but the wreckage quickly sank.

Feb 7 – At 0039 a submarine crash-dived before an attack could be made. The Sunderland loitered and at 0200 an ASV contact was picked up; eight AS bombs were dropped in a dive attack, the last bomb bursting near the stern of the submarine. The submarine turned and dived, leaving a yellow streak. ASV picked up another contact at 0307 and this was attacked with 750 rounds of machine-gun fire.

The citation for the award of the DFC to Wg Cdr Kenneth Garside reads:

This officer has served with the Squadron for the past two and a half years. He has completed over 100 sorties, including attacks on enemy submarines at night. In one of these attacks his aircraft came under enemy machine-gun fire whilst flying at low level. Wg Cdr Garside has performed much valuable work both operationally and in the training of new pilots. (Air Ministry Bulletin 7113)

March and April were somewhat quieter for Sqn Ldr Garside – he recorded two attacks on 27 March. Other members of the squadron were busy in this period, six attacks and three other sightings being recorded. The attacks were made by Flt Lt Milligan (27 March, 23 April and 26 April), Flt Lt Squires (19 April), Flt Lt Brown (22 April) and Flt Lt Frame (28 April).

The hectic period continued into May, four attacks and a number of other sightings being reported. Plt Off Howell (L5806) attacked a submarine on 1 May with bombs and depth charges, the latter falling along the starboard side and causing the U-boat to 'slowly sink with no forward speed'. Flt Lt Frame (W4022) was back in action on 11 May, dropping four AS bombs in an attack that appeared to damage the target. On 26 May it was Wg Cdr Garside's turn again:

At 1927 hours an ASV plot was obtained four miles on starboard beam and found on investigation to be a submarine on the surface. Attack was carried out with four 250lb AS bombs (set at one second delay) with flash bombs attached, and four 250lb depth charges set at 50ft with Mk X pistols. Bombs were observed to hit conning tower of U-boat. After smoke and spray had subsided no further trace of the submarine was seen. It is to be noticed that the flash bombs fitted to the AS bombs assisted greatly in observing result of attack. It is considered that the U-boat was destroyed. Three separate ASV plots of aircraft in close vicinity of attack were picked up immediately attack was completed. General impression being that these aircraft were circling submarine, on protective patrol, for they closed position of attack immediately after flash bombs was seen. Own aircraft continued to take evasive action. (230 Squadron ORB, May 1942)

Two days later Fg Off Howell (L5806) dropped four depth charges on a surfaced U-boat but no result was observed.

Although 202 Squadron operating from Gibraltar was still primarily a Catalina squadron, it did have a number of Sunderlands on strength for much of 1942, these flying ASPs in the Mediterranean and out into the Atlantic. Operational capability was increased by the detachments from other Sunderland squadrons, 10 Squadron RAAF deploying a number of aircraft to Gibraltar, and the occasional attachment of aircraft from other units. These increases were normally connected with specific operations, usually a particular convoy.

No. 230 Squadron recorded a number of attacks in spring, primarily in the waters off Tobruk, but with no confirmed scores, including a number of attacks by Wg Cdr Garside. In one attack on 26 May his crew damaged the Italian submarine *l'Argo*; the submarine abandoned its mission and tried to limp home on the surface. It was found two days later by a 10 Squadron RAAF Sunderland.

On 28 May, Fg Off Pockley of 10 Squadron RAAF, flying W3983/R, was tasked to search for a submarine that had been attacked the previous day by a Catalina:

Sighted oil patch with streak running ENE; saw U-boat at head of streak. Aircraft dived to attack from 1,000ft up stern as U-boat turned sharply to port but was forced to take evading action owing to heavy gunfire, circled U-boat which remained surfaced. At 1415 dived to attack and raked U-boat with machine-gun fire, most of which struck conning tower. No depth charges dropped as run unsatisfactory – large quantity of light and medium flak put up by U-boat. At 1427 attacked from astern and large quantity of flak. As U-boat turned to port four depth charges released from 40ft, overshot and detonated 20–30yd on starboard bow. Aircraft port bomb circuit unserviceable throughout so port depth charges manhandled to starboard bow racks for second attack. U-boat continually opened fire with large calibre gun when approached within 3 miles. At 1642 aircraft attacked up U-boat track as she turned to port, released four depth charges from 30ft which straddled U-boat from port to starboard across conning tower. U-boat completely lost to sight in spray, speed immediately reduced to 3–4kt and course became erratic. Subsequent observations showed large apparent dent on port side just forward of conning tower near waterline. Aircraft sustained several hits, considerable trouble experienced with aircraft machine-guns during action, three of the four tail guns being u/s. Nose gun went u/s during first attack. Great need was felt throughout entire action for forward-firing cannon as extremely heavy damage could have been inflicted on U-boat during and after above attacks. At 1825 aircraft led a Hudson to the U-boat.

The submarine made it back to port two days later.

6 June saw the Australian squadron in action again, Flt Lt Marks dropping eight depth charges on a U-boat, probably the Italian submarine *Brin*; although no damage was inflicted to the submarine it did manage to hit the Sunderland in the starboard engine and float. The following day it was 202 Squadron's turn to find and attack a U-boat, the crew dropping a perfect straddle on their first run. Another four depth charges were dropped as the U-boat submerged. The most likely victim was the *Veniero* as Italian records show that this submarine went missing at around this time. 13 June saw another attack by a 10 Squadron crew, Sqn Ldr Burrage dropping depth charges on a surfaced submarine (both the *Bronzo* and the *Otario* claimed to have been subjected to air attack this day; most records credit the aircraft with damaging the latter). An

EJ155 on trestles at Marsaxlokk Bay, Malta; the aircraft was destined for 88 Squadron at Kai Tak.
Ken Delve Collection

account of this attack was included in the *Coastal Command Review*:

> At 0832, Sunderland O/202, from Gibraltar, flying at 4,000ft sighted a submarine 9 miles ahead, making 12kt. The aircraft dived and attacked on the beam, releasing seven depth charges, set to 25ft, one hung up. They straddled the submarine forward, and the explosion obscured it completely. When seen again, it appeared to be listing heavily to starboard and slightly up by the stern, but righted itself immediately. A thick stream of brown smoke poured out behind the conning tower for the next half-hour. As the aircraft circled, the submarine kept its bows always turned toward it and fired the forward gun. At 0950 the remaining depth charge had been changed to a serviceable rack and the aircraft was turning to make another attack when the submarine was suddenly lost to sight.

Plt Off Egerton of 10 Squadron RAAF was engaged by a Ju 88 on 14 July during which combat the flying boat gunners scored a number of hits leaving the Ju 88 diving towards the sea emitting white smoke. The following day Plt Off Lawrence was on a cross-over ASP when he saw a Cant 501 land on the water and burst into flames, the crew of six jumping into the sea. The Sunderland landed and picked up the crew; after returning to Gibraltar the Italians were entertained in the Mess until the early evening when they were taken to the Spanish frontier.

One consistent theme in the records for 1942 is that of engine trouble; indeed, this is a regular comment from many of the Sunderland squadrons and the Pegasus engine appears to have been a consistent problem, although the only warning note in the Pilots Notes was to avoid rpm between 2,250 and 2,550 because of vibration.

The latter part of August was a poor one for 202 Squadron with the loss of a number of aircraft: W6003 capsized and sank after landing at Gibraltar on 12 August, all the crew escaping without injury. On 28 August, W4029 landed after an escort sortie and burst into flames; two of the depth charges exploded and only one of the pilots survived the incident. Two days later, all four engines cut-out on 'R' due to an air lock in the fuel; however, the aircraft landed safely and was taken in tow.

Another U-boat was attacked by Fg Off Walshe (W6002) on 14 September. Two of the five depth charges exploded against the port bow and two by the conning tower, lifting the submarine out of the water. While the Sunderland circled the scene, the submarine crew abandoned the stricken vessel, the *Alabastro*. This was 202 Squadron's one and only U-boat victory with the Sunderland: the squadron bade farewell to its Sunderland element in October.

The most significant event in the Western Mediterranean in late 1942 was, of course, Operation *Torch*, the Allied landings in Tunisia. There was great concern amongst the planners that enemy naval vessels, including submarines, would play havoc with the vulnerable invasion fleet. A major air component was built into the plan, the Sunderlands from Gibraltar playing a very small but nonetheless important part

On 9 January 1943, No. 230 Squadron moved to Dar es Salaam; they did, however, continue to provide detachments to the Middle East when required. 'Following completion of the North African campaign and the possible invasion of Sicily and Italy, six flying-boats of the Squadron were ordered to proceed to Aboukir and await further instructions' (230 Squadron ORB, June 1943). The first aircraft, DP180 (Wg Cdr Taylor), left the same day, making night stops at Kisumu and Khartoum en route to Aboukir. On 11 June the Aboukir detachment was ordered to Bizerta under command of No. 216 Group Advanced, the primary task being transport work between Bizerta and Malta. The aircraft were also tasked with ASR work on behalf of No. 242 Group.

The ASR task soon produced results:

> Sunderland JM659/Q, Fg Off D. McNichol, was ordered to search for two dinghies containing survivors, alight and pick them up if possible. Took off in path of full moon. SE [Special Equipment] became unserviceable through interference and on approaching Maritimo Island two E-Boats were identified by flashes from their exhausts. Nothing was seen in the positions given but a search was made of the immediate area and a dinghy was sighted in position 4020N 1300E. It was identified by torchlight flashes and a flame float flare path was dropped preparatory to making a landing. At this stage an enemy aircraft came into view and as the Sunderland's mid-upper turret was out of action it was considered inadvisable to alight and course was set for base. (230 Squadron ORB, 17 July 1943)

At 0548 the following day, Fg Off G. Watson was airborne from Bizerta in EJ141 to locate the dinghy:

> An escort of fifteen P-38 Lightnings was provided. The dinghy was sighted 15 miles downwind of the position given by 'Q' [80 miles from Naples] and was clearly identified at two miles distance. Aircraft alighted and took off the complete crew of six belonging to a Marauder. Three of the crew were uninjured, two of the remainder had minor injuries and the third a broken arm and leg. After take-off the aircraft set course for a position 40 miles south of Naples to try and locate another crew down in the sea but after a brief search course was set for base on account of the limited endurance of the fighter escort. Excellent co-operation was given by the fighter escort enabling the crew of the Sunderland to devote their full attention to the rescue work. (230 Squadron ORB)

A second aircraft was airborne that afternoon to search the same area, an empty dinghy was located but the sortie was not a complete waste as the escorting Lightnings downed fifteen Ju 52s that were unfortunate enough to be in the same area.

An uneventful ASR mission on the 19th was followed by a night sortie flown by Fg Off Statham, EJ143 taking-off at 2050 to search an area off the coast of Sardinia:

> Commenced a square search at 500ft and on sixth leg at 2242 hours, observed a red flare fired from a Verey pistol. A dinghy was distinguished and a flame float dropped, four more flares were dropped to make a flare path and aircraft alighted in a heavy swell of 7–8ft. The dinghy was still firing off flares and appeared to be about one mile ahead of flare path. It was thought there was a second dinghy downwind of the first one but on investigation it was found to be a flame float. Searching, therefore, continued for the original dinghy. Numerous Verey cartridges were fired off and the Aldis Lamp was used but no answering reply could be obtained and so touch was lost with the dinghy. Although aircraft was being pounded by the heavy swell, it was decided to remain waterborne and endeavour to pick up the dinghy visually at dawn. Aircraft was waterborne for three hours during which time most of the crew suffered from seasickness. Nevertheless, the Aldis was used frequently and navigation lights were kept on in the hope of getting some reply from the dinghy crew. German voices were heard on the intercom and from the first glimpse of the dinghy, which appeared to be of oval design, the conclusion was formed that it was an enemy crew. As dawn approached and there were no further developments, aircraft took off in a 10ft swell with sea breaking right over it. (230 Squadron ORB)

Other ASR missions were flown during the month, but the above examples demonstrate the determination of the Sunderland crews. This determination was appreciated, as witnessed by a letter from Col Karl E. Baumeister, CO of the US 320th Bomb Group:

> This organization appreciates the wonderful effort put forth in the rescue of the crew we had down in the sea. This rescue has done more for the morale of this organization than one can realize. A service of this type goes a long way in the minds of men flying on missions.

The Bizerta detachment returned to Aboukir at the beginning of August; the Sunderlands now spent most of their time flying passengers and freight, much of the latter being mail, around the Mediterranean theatre, although ASR and other sorties were also flown as required. The flying boats were often operating into hazardous locations:

> Fg Off Statham in W4021 flew from Aboukir to Castellorosso on November 7 with freight and passengers. After unloading, a motor launch hit the aircraft and put one of the port engines out of action. The Captain tried a three-engined take-off but hit an uncharted reef and the aircraft started to sink. To avoid the possible chance of salvage by the enemy in the event of his capturing the island, the Captain taxied the aircraft out in to deep water where it sank in 60 fathoms.

The ending of the 230 Squadron detachment brought Sunderland involvement in the Mediterranean/Middle East to a virtual end, although aircraft still used Gibraltar and the other anchorages from time to time. The numbers of aircraft in this theatre had never been large but the Sunderland achieved an excellent record.

Although no Sunderland squadrons were based in the Middle East in the post-war period, a number of units acquired aircraft for communications purposes. The Iraq Communications Flight used a few Sunderlands – EK595 is seen here being scuttled on 1 April 1946. Andy Thomas Collection

CHAPTER SIX

West Africa

By spring 1941 Coastal Command's area of coverage had extended down to West Africa with the establishment of a Sunderland unit in Sierra Leone.

A single U-boat had made its presence felt off the coast of West Africa in July 1940, one ship being sunk in July and one in August. The threat increased in the latter part of the year with four ships being lost in November and five in December, all within 600 miles (1,000km) of Freetown. As the shipping lanes around West Africa were important to the British war effort, it was decided to station a Sunderland squadron in the area.

On 15 January 1941, No. 210 Squadron received a warning order to detach three aircraft, with crews, maintenance personnel and spares, to Freetown, Sierra Leone for operations; P9633, L2163 and T9041 were duly prepared and departed Oban the following day for Pembroke Dock. The same day, this detached flight was declared to be 95 Squadron and the aircraft, with N9027 taking the place of L2163 as it had better engine hours, were given an additional 355gal (1,600ltr) petrol tank in the bomb compartment, bringing the total to 2,400gal (11,000ltr). The first aircraft arrived at Gibraltar on the last day of January, and remained there for some days, flying escort sorties. Meanwhile, the ground party left Oban and boarded the SS *Highland Brigade* at Glasgow, eventually arriving in Freetown on 2 March. The first aircraft, L5805, arrived via Bathurst on the 17th; not all the Sunderlands made it, however: Sqn Ldr Lombard in P9623 had to force-land in Portugal when he ran short of petrol, aircraft and crew being interned; the crew were released on 25 March but without the aircraft. A combination of severe weather and stronger than expected headwinds had led to the force-landing, P9623 was only slightly damaged and the aircraft was subsequently repaired. With its camouflage removed, and wearing the serial '136', it was put into service with the Portuguese Navy on transport duties to the Azores.

The first operational sortie out of Bathurst was flown by N9050 (Fg Off Baggott) on 24 March, escorting convoy SL69. A total of eight operational sorties were flown in March. The squadron came under the command of C-in-C South Atlantic, whose HQ was at Freetown.

Having arrived in West Africa, 95 Squadron found the situation at Fourah Bay (often recorded as Fura Bay), Freetown, far

Bathurst early 1942, 204 Squadron's T9072 in an atmospheric shot. Peter Green Collection

from ideal. In a report dated 20 April, the CO, Wg Cdr Pearce wrote:

> Contrary to my expectations, I found on arrival at Freetown on 5 April that the present operational location for Sunderland aircraft was in Fourah Bay, where three buoys in reasonably sheltered water are available in the form of a trot. Conditions are, however, generally unsuitable for the operation of flying boats from Fourah, particularly in respect of night flying. A slight swell is prevalent in the open roadstead and this is aggravated by a very strong tide which can reach 5kt.
>
> The Bunce River site is suitable for operation of flying boats but it is impracticable for aircraft to taxy 6 miles in this climate on account of overheating.
>
> Accommodation at Fourah Bay College is both primitive and inadequate. At present the vast majority are accommodated in tents, such accommodation in a place well-known to be malarial constitutes a serious menace to health.
>
> Bathurst presents an ideal site – almost unlimited area of sheltered water with no mountains or obstacles in any direction, the climate is more favourable and the accommodation better. Furthermore, a slipway exists that could be modified for Sunderlands.

German U-boats began operating off West Africa in mid-1940; early the following year it was decided to deploy a Sunderland squadron to the area – 95 Squadron arrived in March. Peter Green Collection

A Coastal Command booklet also highlighted the 'far from ideal' conditions:

> The crews and ground staff lived in tents until one day these were washed away. They found refuge in a church. In the rainy season water penetrates everywhere, even through the hulls of flying boats. Malaria and mosquitoes are ever-present and eager foes. The Sunderlands began the work of convoy patrol immediately. They were badly needed. The main difficulty in those days was to obtain sufficiently quantity of spare parts to keep the boats serviced. Much ingenuity was shown. Oil for the hydraulic gear was obtained from ground nuts; packing-box nails took the place of split pins, toilet paper for oil filters. (Coastal Command)

Fg Off Baggott (N9050) was on convoy escort for SL71 on 10 April 1941 and as the uneventful sortie continued he had to shut down the starboard outer engine as petrol was becoming short. At 1820 the starboard tanks were empty and the Sunderland was force-landed two miles off shore; unfortunately the anchor flukes broke and the aircraft had to be beached on the foreshore with the aid of local troops. The aircraft was refuelled overnight from lorries and at 0600 the next day it was hauled off the beach and after an 8-mile (13km) taxy in a moderate swell managed to take-off in fairly calm water. Sunderland N9050 was kept airborne for 40 minutes while minor temporary repairs were carried out to small holes in the after-bilge compartment. The aircraft landed at Bathurst at 0900.

April was a fairly typical month for 95 Squadron:

Date	*Aircraft*	*Captain*	*Time*	*Details*
1 Apr	L5805	Plt Off Bailey	0805–1650	C/V escort
2 Apr	L5805	Plt Off Bailey	0825–1715	C/V escort
5 Apr	N9050	Fg Off Baggott	1340–1755	AS search
6 Apr	N9050	Fg Off Baggott	0730–1700	AS search
7 Apr	N9050	Flt Lt Gibson	1000–1955	AS search
8 Apr	L5805	Plt Off Bailey	0410–1617	AS search
9 Apr	T9078	Flt Lt Gibson	0615–1655	C/V escort
10 Apr	N9050	Fg Off Baggott	0630–1820	C/C escort
12 Apr	L5805	Fg Off Robertson	0410–1550	AS search
16 Apr	L5805	Plt Off Bailey	0445–0945	AS search
17 Apr	T9078	Flt Lt Gibson	0758–1800	AS search
18 Apr	L5805	Sqn Ldr Lombard	0335–1330	C/V escort
19 Apr	L5805	Plt Off Bailey	0415–1400	C/V escort
20 Apr	L5805	Sqn Ldr Lombard	0530–1520	C/V escort
22 Apr	T9078	Sqn Ldr Bisdee	0531–1525	C/V escort
24 Apr	L5805	Sqn Ldr Bisdee	0520–1525	AS search
27 Apr	L5805	Plt Off Bailey	0435–1450	C/V escort
28 Apr	N9050	Fg Off Baggott	0550–1605	AS search
28 Apr	T9040	Flt Lt Evison	1412–2130	AS search
29 Apr	T9078	Sqn Ldr Bisdee	0620–1722	C/V escort
30 Apr	T9078	Flt Lt Gibson	0405–1440	C/V escort

The view from the back. Rear gunner of 204 Squadron's N9024 on 11 November 1941, monitoring two lifeboats from the MS Peru.
204 Squadron Records

At the end of June 228 Squadron began the move to West Africa that had first been promulgated at the beginning of the month, Wg Cdr Brooks leaving for Khartoum on the 23rd on the first leg of the journey (the ground party having left the previous week for Port Tewfik to board the HMT *Dumana*). The Sunderland was to route through Khartoum, Kampala, Stanleyville, Leopoldsville, Libreville, Lagos and Freetown. Engine problems dogged L5803: a port engine was replaced at Kampala but when the aircraft reached Lagos on 7 July, both starboard engines were unserviceable. Engine stands had to be manufactured and a barge hired so that work could be carried out, new engines arriving on a Bristol Bombay transport on 20 July. By this time the CO had received news that 228 Squadron would be replaced at Bathurst by 204 Squadron when the latter had completed overhaul and preparation at Pembroke Dock. Personnel of 228 Squadron due for relief would be posted to the UK whilst the remainder would be absorbed into 95 Squadron, along with the aircraft. Sunderland L5803 eventually made it to Freetown on 28 July; meanwhile, T9046 was also finding the journey difficult, the crew discovering at Lagos that a blade was split and holed:

> As there was no replacement LAC Bindong started work on the old airscrew, filing down cracks and tips. There a was large chip in one blade so a corresponding one was made in the other two blades. There was no balancing machine available so the engine was run up to test for vibration – none was experienced. The airscrew was changed with the starboard outer airscrew to avoid further damage and the aircraft tested and found serviceable. The repaired airscrew was not subsequently changed until the aircraft was flown to England for major inspection. (228 Squadron ORB, July 1941)

Major overhauls were always carried out in the UK and for this trip aircraft were fitted with overload tanks.

No. 228 Squadron's stay in West Africa was indeed short-lived, the unit returning to the UK in late August 1941. The first four of 204 Squadron's Sunderlands (W3981, W3980, T9044 and T9074) had left Pembroke Dock for Gibraltar on 15 July, remaining there for a few weeks during which they flew convoy escort and ASP. More aircraft arrived at Gibraltar during July, although two deployed to Malta at the end of the month. However, on 27 August the CO, Wg Cdr Coote, flew T9074 to Bathurst, moving on to Freetown the next day. By mid-August the squadron had most of its aircraft in West Africa, with an average of four aircraft serviceable at Bathurst and one at Freetown.

The routine of ASPs and convoy escorts was soon established, although operations were never very intense. There was, however, opposition from French Vichy forces and two of the squadron's Sunderlands were attacked by fighters on 29 September. 11 October saw Flt Lt Ennis in difficulty following a convoy patrol. The aircraft was forced north by a severe storm and was eventually forced to land 5 miles (8km) south of Dakar, out of petrol. The following day the CO was airborne, searching for T9074 and managed to land alongside, transferring enough fuel for the Sunderland to get airborne and return to base.

In February 1942 two U-boats, U-68 and U-505, were sent to operate off West Africa; the latter was attacked by a Hudson on 28 March and a Sunderland on 18 April, but with no apparent result. The Sunderland attack was made by Flt Lt Dart

(204 Squadron); Flt Lt Dagg sighted a periscope the following day but was unable to make an attack.

However, one of the consistent threads of the 204 Squadron records for this period is that of search and rescue, for crews of both ships and aircraft. On 10 June a Sunderland was tasked to locate a downed Hurricane pilot: the aircraft found the dinghy, landed alongside and rescued the pilot. At the end of the month the squadron was searching for one of its own aircraft, T9041 (Flt Lt Ennis) having failed to return from a convoy escort. An air search over the next two days proved fruitless but a message was then received that ten of the twelve crew had been rescued by HMS *Velox*. The loss of L2158 (Plt Off Quinn) on 17 August had an equally good outcome, the survivors being picked up a few days after the aircraft had been lost. Sunderland T9070 was destroyed at its moorings at Half Die when it caught fire and the depth charges exploded.

'Flying boat under full sail – one knot'; T9074 took to the water, short of petrol, on 10 October 1941. Ken Delve Collection

There were now five submarines operating in the area by July and these sank seven ships in the last ten days of the month. This strength was increased under Operation *Eisbaer* with four U-boats tasked to attack shipping off South Africa. In the last week of August all available U-boats in the area were tasked against convoy SL119, sinking five ships in the first week of September (in addition to successes off South Africa). Whilst success eluded the Sunderlands, there were losses. The instances of Sunderlands catching fire at their moorings continued into 1942: on 22 August, L5803 of 95 Squadron was 'blown to pieces, the result of a depth charge exploding following a fire which started on the refueller moored alongside. The aircraft had just flown an ASP of the Southern Shipping Lane.' There does not appear to have been a common cause to the various Sunderland losses of this nature, or at least not one that has been recorded in official documents.

3 February 1941 and 204 Squadron locates a downed Ensign. 204 Squadron

Sunderland DV975/H of 95 Squadron, flown by Flt Lt Bocock, crashed on 28 November in the River Bunce, off Jui, having been taking part in an exercise with a High Speed Launch; seven of the Sunderland's crew were killed. The remains of the aircraft were subsequently blown up as they posed a danger to navigation.

Many hours of convoy patrol and anti-submarine searches were being flown but the only sorties to achieve any concrete results were often those with more of a mercy mission aspect – searches for downed aircraft. 15 October, for example, saw one of the squadron's aircraft (DV959, Flt Lt Hewlings) locate and rescue five survivors from Wellington HX636.

Operation *Torch*, the Allied invasion of North Africa, at the beginning of November involved the Sunderland units in intensive operations of the MAX areas, and the associated RING and SUN areas, although no enemy submarines were sighted.

In January 1943 it was reported that German submarines were operating off Port Etienne; to counter this, West Africa Command commenced operations from Dakar and Port Etienne, the latter becoming a regular base for Sunderland detachments whilst the former, later in the war, housed 343 French Naval Squadron.

It had been decided in February 1943 to allocate Sunderlands to at least one Free French unit in West Africa, the first such re-equipment, comprising two aircraft (JM674 and DP182), being allocated to the Bel Air, Dakar-based 4th GR Squadron. Additional aircraft were delivered in the summer and Sunderlands equipped 4E of 1st *Flotille*, this unit being re-designated 7th GR *Flotille* (7FE) in October. However, as part of West Africa Command the unit was also given the RAF designation of 343 Squadron. This unit was formed from French Air Force units

This photo shows the rough and ready nature of some of the bases in West Africa Command. Ken Delve Collection

'Sunderland base West Africa' – a wartime graphic designed to give nothing away. Ken Delve Collection

following the success of the Allied landings in French North Africa, its first Sunderlands being allocated in the summer; the unit remained operational throughout the war. According to the 204 Squadron ORB for March 1943 'The French Sunderland, DD833/M undertook its first operational sweep with a French crew under the command of Lt L.V. Giraud. The aircraft took-off from Half-Die and landed at Port Etienne. ... the second French contingent arrived at the end of the month for Sunderland training' (204 Squadron ORB).

By March the majority of sorties by 95 Squadron were being flown from Half Die, most of these being convoy escort or air-sea rescue. This was a typical month, with thirty-two operational sorties being flown: fourteen convoy escorts, four ASPs, eight reconnaissance sorties and six searches. On 31 March, W6063 was involved in a search for survivors from a merchant vessel when the aircraft, for no obvious reason, crashed into the sea; there were only three survivors.

There was still the occasional contact with submarines. Operating from Port Etienne on 5 April, N9024 (Fg Off Copper) was on a Cape Verde patrol when it was diverted to escort convoy SL127. At 1600 hours the crew spotted a 500-ton U-boat on the surface and made a gliding attack. The Sunderland's approach appeared to be unobserved until about 300yd (270m) when the submarine's machine-guns opened fire, only to be silenced by the flying boat's front turret guns. Six depth charges were released and one scored a near miss, the submarine being seen to roll. The aircraft attacked again, this time straddling the stern with two depth charges. A third attack was made with machine-guns, by which time the U-boat had submerged. The Sunderland was credited with a damage. 'A vivid green patch with a white ring was observed, but no wreckage.'

On the same day, Flt Lt Hugall and crew undertook a search for a missing USAAF Liberator; they found two dinghies and landed alongside, picking up the ten American airmen. Sadly, the month ended with the loss of one of the squadron's Sunderlands ('A'); the aircraft was in transit from Gibraltar to Bathurst when it had to force-land at night south of Port Etienne. Subsequent air searches found wreckage but no survivors.

Another aircraft was lost the following month, JM680/J hitting 'H' of 95 Squadron on take-off from Half-Die on 31 May, the impact tearing a large hole in the Sunderland's hull. After circling for some time, dumping fuel and jettisoning depth charges, the aircraft was put down just off the end of the slipway. Although it sank in seconds, all the crew escaped safely. The 95 Squadron ORB recorded the same incident: 'At 0415 on May 31, W6062/H was sitting on its mooring when it was struck by a 204 Squadron aircraft, causing extensive damage to the port wing, midships gun turret and tail fin.'

Sunderland JM671/Z of 95 Squadron was on a flight test on 22 May when

> ... the port outer engine failed due to an oil leak. The aircraft set course for base on two engines [not recorded as to why only two engines], and approximately 3 miles from harbour mouth in bad visibility the port inner also failed due to an oil leak. The Sunderland landed cross-wind in the Bay to avoid running on to the coast. The aircraft was undamaged and was towed in.

Having undertaken repairs, a test flight was flown two days later and all seemed to be OK until the second take-off, when 'smoke came from the port inner engine and filled the flight deck. The aircraft immediately landed straight ahead; having suffered no further damage it was towed in. The fire was due to the generator wiring burning out.' The following day the Sunderland was tasked for a convoy escort and the engines appeared in good order on the run-up test; however, the starboard inner cut-out immediately after take-off – with the aircraft at only 50ft there was no time to jettison fuel and it had to be landed straight ahead with a full load of fuel and depth charges, as well as ten crew: 'A slight swerve was made to starboard on the step to avoid a native canoe, and owing to this swerve the plates of the starboard side of the hull were strained.' Engine problems continued to plague 95 Squadron (and indeed are a recurrent theme throughout the Sunderland's career), although there does not appear to have been any consistent cause and no engineering defect notifications were raised.

May brought an increase in Sunderland strength when 490 Squadron RNZAF at Jui gave up its Catalina IBs in favour of Sunderland IIIs, the first operation being flown on the 17th.

15 October 1942 and the rescue of a Wellington crew by Flt Lt Hewlings (204 Squadron); five survivors were picked up off the beach. Ken Delve Collection

'KG-V' on the stand at Half Die 1943, whilst airmen take advantage of the Naafi 'tea and wad' truck. Ken Delve Collection

West Africa, East Africa and the Indian Ocean

The scene of activity was, however, to transfer to East Africa and although operations from this area could be seen as part of the 'Indian Ocean campaign', they have been included here as part of the African chapter.

With the increasing threat of submarine activity in the western part of the Indian Ocean, and more especially the shipping lanes around Madagascar, it was decided to strengthen the anti-submarine forces in East Africa. No. 230 Squadron had begun its move to East Africa on 9 January 1943, Flt Lt P. Squires in EJ136 departing Aboukir for Wadi Halfa on the first leg of the trip to Dar es Salaam, Tanganyika. The squadron was in place at its new base by the end of the month, but in the meantime, EJ136 had left (17 January) on an Indian Oceans Islands tour, the passengers including Lt-Gen Sir William Platt, GOC-in-C East African Command and AVM Wigglesworth, AOCV East Africa. The tour took in Pamanzi, Diego Suarez, Mauritius, Rodriguez, Seychelles and Mombasa. In addition to giving the senior officers the opportunity to visit locations within their vast operational area, the tour also gave the Sunderland crew an opportunity to survey possible anchorages for future use. The Indian Ocean islands tour became a regular part of the squadron's routine. It was not until 25 March that the main party arrived aboard the SS *Takliwa*. Having left the relatively healthy climate of the Middle East, the squadron was dismayed by its its new home: 'the Mosquito authorities inspected the Squadron camp and found evidence of Yellow Fever and Malaria.' The problem of malaria was to task the squadron for the duration of its stay in East Africa, at one point the squadron recorded that an epidemic had broken out, ninety-four cases being reported (May 1943). By 13 May the squadron had twelve Sunderland IIIs on strength but it was still having difficulties with facilities:

> A decision was made to complete work on the slipway under construction at Kurisini Camp but close down work on the Flying Boat base and work was therefore suspended. A few weeks later the decision was made to continue with the Flying Boat base and so work commenced immediately. (230 Squadron ORB, May 1943)

Throughout its operational history the Sunderland had been subject to official and unofficial armament modifications; as CO of 230 Squadron at Dar es Salaam, Dundas Bednal was determined to provide his aircraft with a more effective forward-firing gun.

> I had a talk with the Armament Officer and we though that the best way to deal with this menace [U-boats fighting back] was to try and match the U-boat's firepower and range by installing a 20mm cannon in the bow compartment of a Sunderland. We just happened to have a new 20mm cannon in our stores. It was to have been a fixed gun to be fired by the pilot and aimed by means of an open 'ring and bead' sight. The Sunderland seemed quite robust enough to stand the strain imposed by firing the gun. Unfortunately this unofficial modification was discovered during a Command inspection and was promptly stopped. (Dundas Bednal)

August saw four aircraft of 230 Squadron detached to Pamanzi to concentrate effort against U-boats believed to be operating in the Mozambique Channel, much of the task involving providing escort for naval forces. One such mission was flown on 20 August by Flt Lt A. Todd in EH131/T:

> Aircraft failed to locate the naval force after 5½ hours' flying and sent message to base to this effect. The dead reckoning position of the convoy was sent to the aircraft at 1400 hours with instructions to continue the sortie until 'prudent limit of endurance' and then to proceed to Tulear. A message was received from the aircraft at 1450 hours that it had arrived at the position given but that there still no sign of the convoy. At 1838 hours aircraft appeared to be in difficulties as a message was sent to Tulear asking if they had heard engines overhead and a request was made a few minutes later for bearings from Dar es Salaam and Diego Suarez. In response to this request, first class bearings were given and from 1831 hours continuous messages were sent from base instructing the aircraft to steer due east and pick up the coast. At 1906 hours aircraft asked Tulear for D/F assistance and at 2003 sent an SOS that it was making a forced landing. No further messages were received thereafter.

Despite subsequent searches by a number of Catalinas, no trace of the Sunderland was found.

The Indian Ocean islands tour was now a very regular event; indeed, four such were flown in September 1943. For 95 Squadron the latter part of the year brought poor serviceability at both Port Etienne and Jui; only two aircraft being serviceable each day, although this proved adequate to meet the low level of operational commitment.

In what was probably the most tragic week for any of the Sunderland squadrons, 204 Squadron lost three aircraft in the space of ten days. Fg Off J. Finney (JM710/L) was returning from an ASP at 2100:

> At 2110 the aircraft was seen by Fg Off Watson, on duty as Flare Path Officer, to pass over the flare path. He immediately gave them a green, and went below to put the Aldis Lamp away. On coming on deck he saw the aircraft dive straight into the sea, bursting into flames. It is presumed that the aircraft stalled in on a left-hand turn from a height of approximately 800ft. At the time of the accident there was no moon, and a slight drizzle had just commenced. (204 Squadron ORB)

On 1 October Flt Lt C. Mayberry (DV975/G) crashed carrying out a night landing at Bathurst, seven of the crew being rescued by Marine Craft. The following day, W6079/C crashed whilst trying to take-off from Port Etienne: 'After two unsuccessful attempts, on the third attempt the aircraft swung to port and crashed on the beach. The navigator, Fg Off Dunn died of his injuries.'

According to the 230 Squadron ORB for December,

> Operational flights were made in close co-operation with Flag Officer East Africa, No. 246 Wing and with other squadrons in the area. Forces were disposed to meet a possible U-boat threat in (a) the Mozambique Channel and (b) the Mombasa area. Aircraft of the Squadron flew three sorties on A/S sweeps in the Mombasa area and a similar sortie was flown in the Mafia Channel area. The aircraft carrying out the latter was, at the time, on a transit flight and was switched over to the more pressing operation – an incident which shows the versatility of the Squadron. Transport flights were many and varied, ranging from short trips between bases along the East African coast to long communication flights between the Indian Ocean Islands bases and transportation flights between Kisumu on Lake Victoria, Nyanza and Khartoum. In spite of many difficulties which were more often than not traced to engine trouble, all the work of this nature was satisfactorily completed. Aircraft of the Squadron performed a useful job of work in transporting u/t aircrews from Kisumu to Khartoum, en route to the Middle East. For some days a regular shuttle service was operated and this relieved what was tending to become a congestion of personnel at this stage of their journey. The month was marred by the loss of EJ140/V with her entire crew. The aircraft took off from Mombasa on December 29 to carry out M/F D/F calibrations and at approximately 0625 crashed into the mountains in the Voi region. It was thought that the aircraft was flying at the time through heavy clouds. (230 Squadron ORB)

AVM J. Cole-Hamilton, AOC West Africa, issued an instruction in November 1943 for a

> ... revised establishment for RAF Station Bathurst, placing the servicing of 95 and 204 Squadrons on a centralized basis is now being prepared for submission to Air Ministry. This establishment will provide for all servicing at Bathurst being organized on a maintenance wing basis under a chief technical officer. It is proposed to bring this new establishment into force on January 1, 1944. In the meantime it is important that any re-allocation of technical accommodation should be such that it will fall into place in the new organisation and all sections should be centralized now as far as possible. This applies particularly to electrical and instrument maintenance. There is to be only one Electrical and Instrument Workshop and Store wherein all special equipment and spares are to be centralized. All repair, calibration and test work is to be carried out in this shop except that which the Electrical Officer considers it desirable to carry out on the aircraft. (95 Squadron ORB, November 1943)

A shark's-eye view of a 204 Squadron aircraft at Port Etienne. Ken Delve Collection

This does not appear to have come into effect until March 1944.

On 22 December a 95 Squadron Sunderland departed Jui for Abidjan, flying an anti-submarine sweep en route:

> A small chart of Abidjan had been obtained, also information concerning landing, moorings, marine service, station, etc. When landing at Abidjan, most flying boats use N-S waterway, regardless of wind. The take-off and landing run are over one mile in length, width of waterway not unlike Jui. There are four buoys, one refueller very slow in pumping, one power boat and one dinghy. The Commanding Officer is of the rank of Squadron Leader and sixteen airmen make up the strength of the Station. It is advised that wireless telegraph messages be sent to Takoradi. Feeding is a difficult problem, aircrew dining at French restaurants, the cost being met by the RAF. Abidjan is a very beautiful town and most enjoyable, apart from the fact that there are no washing facilities. Wine is the universal drink. It is undesirable to go u/s or the engineers will be worked to death, as facilities are non-existent.

These little pen-pictures of the various 'remote' operating stations used by the RAF's flying boats are fascinating.

January 1944 opened badly for 95 Squadron at Port Etienne: on the 5th of the month Sunderland DW105/T took-off for a sortie but returned a few minutes later with smoke issuing from the aircraft – it exploded and crashed into the water near Cansado Point, Fg Off Spinney, the second pilot, being the only survivor. His report stated that shortly after take-off the starboard inner engine caught fire, filling the aircraft with smoke and flames and making it impossible for the pilot to see.

By late January only U-175 was left off West Africa and this boat reported being constantly harried by air patrols; in one incident on 19 January the U-boat was working its way towards a large merchant ship when it was found by Sunderland 'G' of 270 Squadron and forced to dive and break off its attack. The same Sunderland found the U-boat again on 30 January and seriously damaged the German vessel. The departure of this U-boat brought an end to operations in this area for some time, and

No. 270 Squadron operated from Libreville and Apapa. Peter Green Collection

even when they resumed they were not on the same scale. January saw another Sunderland unit in the region, the Catalina-equipped 270 Squadron at Apapa re-equipping with Sunderland IIIs, the intention being to convert three of the squadron's experienced crews to the new type, whilst also receiving six new crews. The first aircraft, 'A', had arrived on 22 December but it was late January before a reasonable number of Sunderlands had arrived. February was spent on conversion flying, although the squadron's first operational sortie was flown by Flt Sgt Wall on 18 February, Sunderland 'A' escorting convoy LTS 11. By spring 1944, 270 Squadron had nine Sunderlands on strength and was using a number of advanced bases, including Libreville, Banana, Point Noivre and Abidjan. The number of operational sorties was limited but the squadron did take part in various training exercises, one comment in respect of Exercise *Filto* being recorded:

> Use of Sunderland aircraft was of the greatest assistance and added an important touch of realism to the exercise by proving to African soldiers that his food could come to him from the air, which he was not inclined to believe before it happened. The bombing attack was also a complete success.

On 28 January 1944, No. 204 Squadron moved its main base to Jui. Its operational record for January (*see* table right) shows both the nature and intensity of sorties.

Escort work occupied most of the operational flying, 330 hours of the 467 hours total in February being taken up by such sorties.

An intriguing entry in the 95 Squadron ORB for 21 March stated that 'Sunderland "C", which is operationally unserviceable, left Bathurst for the UK ... this aircraft carried the twin 0.5in front turret designed and constructed by the Squadron. For information on this fit, Flt Lt D Curran, the Squadron Gunnery Officer, accompanied the aircraft.' There is, sadly, no further information contained in the ORB in respect of this turret, a classic (frustrating) example of the historical limitations of such official records!

Earlier in the month, the CO of sister squadron, 204, Wg Cdr Hawkins, had flown to Fishlake to evaluate its suitability as a diversion landing base; this location features in the records of both squadrons from this point on. After a period of poor serviceability, average aircraft availability being only 40–45 per cent, March brought an improvement to almost 60 per cent. However, the 204 Squadron ORB noted two problems:

> On March 2, 'H' returned early from a sortie because of dropping oil pressure. This was found to be caused by the filter being clogged with carbon. On further investigation it became apparent that oil of incorrect specification had been used. All aircraft which had been refuelled from that same source were grounded and oil systems flushed and refilled. The cause was that the Equipment Branch had decanted oil from

Date	*Aircraft*	*Captain*	*Time*	*Details*
2 Jan	EK580/B	Flt Lt Pare	1350–0310	Escort Cv SL145
3 Jan	EK582/L	Fg Off Hewitt	0435–1705	Escort Cv SL145
	DV965/A	Flt Lt Johnston	1515–0335	Anti-sub search
4 Jan	EK580/B	Flt Lt Pare	1039–2355	Anti-sub search
	EK582/L	Fg Off Hewitt	1655–0405	Escort
5 Jan	DV965/A	Flt Lt Johnston	1630–2359	Escort Cv SL145
6 Jan	DW104/K	Fg Off Pallett	0620–1925	Patrol, blockade run
	EK580/B	Flt Lt Pare	0550–1925	Patrol, blockade run
9 Jan	EK587/J	Flt Lt Brown	0600–1945	Escort Cv OS62
13 Jan	EK580/B	Flt Lt Pare	2325–1300	Escort Cv SL146
14 Jan	DW104/K	Fg Off Pallett	0900–2100	Escort Cv SL146
	EK582/L	Fg Off Hewitt	1815–0525	Escort Cv SL146
15 Jan	JM682/H	Flt Lt Hibberd	0510–1730	Escort Cv SR9
	EK580/B	Flt Lt Pare	1400–0250	Escort Cv OS63
	EK587/J	Flt Sgt Todd	1435–0320	Escort Cv SR9
16 Jan	EK582/L	Fg Off Hewitt	1220–2150	Escort
	DW104/K	Fg Off Pallett	2245–1135	Escort
20 Jan	EK587/J	Wg Cdr Hawkins	1125–1510	AS sweep
22 Jan	EK587/J	Wg Cdr Hawkins	0848–1224	AS sweep
	DV965/A	Flt Lt Johnston	1533–2306	Escort Cv OS64
24 Jan	PP188/C	Flt Lt Foulds	0411–1426	Escort Cv SL147
	JM672/E	Flt Lt Brown	0514–1520	Escort Cv OS64
	EK587/J	Flt Sgt Todd	1120–2115	Escort Cv SL147
	JM682/H	Flt Lt Hibberd	1221–0220	Escort Cv OS64

Framed by the port float, St Mary's Island, Bathurst, 1943. Ken Delve Collection

1 February 1944 and a 230 Squadron aircraft cruises off Dar es Salaam; the squadron had moved to this East Africa base in January 1943. Via Arthur Banks

four-gallon tins into 44-gallon barrels and had not marked the barrels with the Specification Number of the oil they put in. Altogether, seven aircraft were unserviceable for 48 hours. ... On March 22 Sunderland 'F' [DW104] whilst on an exercise experienced one depth charge exploding on impact with the water. No definite reason for this could be given, though the accident was thought to have been caused by (a) pistol being incompletely cocked, or (b) the copper disc was not inserted in the detonator pocket. A subsequent check was carried out on all aircraft, and all pistols were found to be correct. (204 Squadron ORB)

The CO also noted that he was short of crews – having only nine crews for eleven aircraft.

One of the major tasks for the Bathurst Wing in April was Operation *Fillet*, an anti-submarine patrol that called for three or four aircraft to be airborne at the same time each day. AHQ West Africa had ordered 204 Squadron to move to Bathurst, eight aircraft and sixty-five groundcrew making the move on 1 April. 'During Operation *Fillet* flying was very intense and was only maintained by the splendid work of the servicing echelon, who worked day and night to achieve the necessary serviceability.' The squadron ORB went on to make a number of significant general comments :

It is considered that night attack on U-boats would be extremely difficult, if not impracticable, owing to lack of an anti-submarine parachute flare equipped with a suitable delay and capable of burning for a long period when dropped at low altitude. Cases of aircraft returning to base early with engine trouble appear to be slightly on the increase. The re-engining of the Sunderland III with a more reliable motor is felt to be overdue, as also is the fitting of full-feathering propellers.

No. 95 Squadron lost another aircraft in April, DV973/P crashing on landing on the 13th. At the end of May, and after a number of delays, all Sunderland servicing and maintenance was taken over by the Station Servicing Wing at Bathurst.

By May 1944, No. 270 Squadron had said farewell to its last Catalina and the Sunderlands were now operating from Libreville and Apapa. Two aircraft were lost the following month: ML811 (Fg Off Graham) was en route from the UK to SEAC when it crashed near Lakka with the loss of all on board; four days later EK585 caught fire at its mooring and the on-board maintenance crew were unable to put out the fire. The aircraft exploded when a depth charge went up. Subsequent attempts were made to salvage the other depth charges, which it was thought had not exploded, but the zero visibility in the water and the soft sand into which the aircraft had sunk made such recovery impossible.

The Sunderland's ability to land on the water in the event of weather or mechanical difficulty has been recorded a number of times in this book; a classic example occurred on 11 July 1944 when 95 Squadron's JM671/Z (Flt Lt Rowntree) had to divert to Dakar after aborting a Creeping Line Ahead sortie:

> ... due to the thunderstorm risk at base, but when the aircraft arrived at Dakar a storm was about to break and therefore no flare path had been lit. 'Z' attempted to return to base around the storm but petrol was short and because of the storm problems it force-landed 23 miles south of Dakar. Preparations were made for the storm, which was weathered OK. At 0809 the following day 'A' [EJ144, Plt Off Skinner] appeared overhead and at 1105 the ASR boat took the aircraft in tow. The line broke and so the Sunderland took-off and flew to Dakar.

This was not the only aircraft search flown by the squadron that day, Plt Off Scargill took DV956 to search for 343 Squadron's 'F', which was found and to which an escort vessel was directed. The storm had caused a number of aircraft to 'take to the water' and two days later 95 Squadron was involved in the search for 'P' of 490 Squadron. A dinghy, and seven survivors, was located by 'M' of 204 Squadron, this aircraft being relieved on station at 2020 by 'A' of 95 Squadron. Finally, in what had been a rewarding month, 'A' was again involved in a successful search, the survivors from ML855 being found.

Flt Lt Pennell of 270 Squadron reported problems with the flare system during a submarine exercise on 30 November:

> Considerable difficulty was experienced in releasing the flares and occasionally a flare would ignite in the chute. We used the four-flare chute from the rear door. It was found that four flares dropped simultaneously gave no advantage over the single flare. The method of attack used was to release the flare from 800ft at 1¼ miles from the target; this method was found impracticable as the flare from the flares blinded the pilot and it was impossible to see the target until it passed directly underneath. (270 Squadron ORB, November 1944)

As with Sunderland operations in Home Waters, many sorties were now flown at night and whilst the 10cm ASV was now working well, the problem of adequate illumination continued to plague the Sunderland units.

November was another month with no operational hours for 204 Squadron:

> The flying hours this month have been devoted entirely to flying training, in the absence of operational commitments. Exercises *Bauxite* and *Tarpon* were carried out with the co-operation of the Navy, and found to be very instructive. The Squadron has also carried out exercises with an actual submarine, which training is of the highest value. Two aircraft carried out a coastal familiarization trip. Navigation exercises were carried out by second and third pilots during passages between base and base. During exercises *Bauxite* and *Tarpon* parties of air crews were embarked on the ships to observe the action from the naval side. This was found to be both a pleasant change from normal duties and to provide excellent experience. The Navy have expressed their willingness to make available to air crews the 'Submarine Attack Trainer' and to provide the instructors and staff to operate this apparatus. It is hoped that this training will acquaint air crews with the problems of surface vessels subjected to submarine attack, and the limitations of the detecting apparatus employed by these vessels.

By November the operational commitment for the West Africa Command units had, according to 95 Squadron, virtually ceased:

> Since July 1944, operational commitments have gradually fallen off until during November only thirteen operational hours were flown. The Squadron serviceability was high and the reason for the small amount was entirely due to the lack of opportunities. This lack of opportunity was again due to the fact that no U-boat was appreciated to have been in the Command area since about October 11, and consequently no cover has been required for convoys. On or about November 7, convoy OS93 dispersed off Port Etienne and proceeded to Freetown, each ship individually or in small groups, at their most economic speed. This policy has been maintained for all subsequent OS convoys. The only operations carried out by this squadron during November was an ASR and one or two sorties laid on for the purpose of checking on these individual ships as they spread out down the coast. In order to maintain the routine effort of the Squadron, the policy laid down by AHQ West Africa has been complied with and training has been substituted as far as possible for lack of operational hours. This training has taken four separate forms:
>
> 1. More concentrated Second Pilot training in the form of circuits and landings.
> 2. More bombing, air firing and Army/Naval co-operation exercises.
> 3. Introduction of OFEs (Operational Flying Exercises) on the lines of those carried out by 4 OTU, Second Pilot acting as Captain and the normal captain or a third pilot navigating.
> 4. Introduction of training and familiarization cruises:
>   a. Jui–Fishlake–Bathurst.
>   b. Jui–Lagos–Fishlake–Jui–Bathurst.

A new series of patrols – 'Kite' and 'Kestrel' – was instituted by AHQ West Africa in December in case any of the new Type XXI and Type XXIII U-boats attempted to slip into the area. Two aircraft from each squadron, 95 and 343 French Naval, were tasked to fly alternate patrols, although, in the event, the few patrols that were actually flown proved uneventful. In the whole of December, 95 Squadron flew only two operational sorties, both being 'searches for a suspicious light'.

Operational patrols had resumed for 204 Squadron in late December and the standard anti-submarine patrols flown from January 1945 were primarily 'Falcon' and 'Buzzard'. There was no activity to report and the only incidents of note usually involved problems with the aircraft; an unusual, and potentially lethal, incident occurred on 13 January: Flt Lt Warrior was one minute from Aberdeen Point in Sunderland JM682/H when

> ... the starboard float main spar snapped, the slipstream blew the float into the ailerons, jamming the controls. The aircraft dropped from 1,200ft to 500ft before any control was obtained. This was only possible with the aid of all three pilots. The aircraft did a very wide circle to starboard to return to base, the depth charges and 1,000gal of fuel being jettisoned. The aircraft landed safely with the crew in ditching positions, and then they went to the port wing to maintain stability.

This was indeed a fine piece of flying and crew co-operation.

The 'Buzzard' and 'Falcon' patrols continued into February, although the squadron made preparations to

> ... change over the 'Kittens' and 'Camels'. The introduction of the latter would have involved landing at Bathurst overnight and would have given the Squadron an opportunity of reaching the new operational commitment of 444 operational hours. Before they could be commenced, however, orders were received to prepare for an intensive U-boat hunt. Nine days were spent in these preparations but unfortunately this time was wasted as the hunt failed to materialize. (204 Squadron ORB)

This was later reinstated, the squadron mounting intensive patrols from Fishlake, Jui, Bathurst, Dakar and Port Etienne for a northbound U-boat thought to be between Ascencion Island and Cape Palmas.

Catalina and Sunderland at Dar-es-Salaam. Arthur Banks

Numerous ASV contacts were made but there were no sightings; nevertheless, it did provide crews with an opportunity to refresh their skills in what was still their prime task. The high level of operations in March led to a reduction in hours during April as a shortage of Pegasus engines meant that a number of aircraft were unserviceable. A new patrol area, 'Dog', took over from 'Kittens' and 'Camels' in April, three aircraft being tasked every other day with these 10-hour barrier patrols. These were flown in conjunction with 490 Squadron, each squadron providing aircraft – two one day, and one the next in rotation.

Sqn Ldr Gall was called-out on 1 March to conduct an ASR search for a missing Mosquito; the aircraft was found at 1135 and the Sunderland dropped two Bircham Barrels and a dinghy pack. At 1317 it landed on the Cacheo River and a rescue party was sent ashore to burn the Mosquito and return with its crew. The following month it was the turn of a 95 Squadron crew to need rescue, DV956 having to force-land after engine failure, Flt Lt Wood and crew being picked up by the Portuguese ship *Guine* and landed on the Cape Verde islands.

Meanwhile, on 3 March, the first Sunderland Vs arrived at Dar es Salaam from Kipevu as re-equipment for 259 Squadron, this Catalina unit having operated from this base since the previous September. Ten days later, with three Sunderlands now on strength, one aircraft was sent to Mombasa to commence crew conversion training. By late March three Sunderlands were at Mombasa on conversion duty whilst further aircraft, and 'major stocks from Blackburn', arrived at Dar es Salaam. 'Training continued throughout the month and has been performed with keenness. All crews

Over Mombasa. Wg Cdr Woodward

have shown an earnest desire to adapt themselves as quickly as possible to the new aircraft. By the end of the month three crews had been fully converted and two for day only.' However, the squadron then came across a problem with the aircraft:

> Operational training has been temporarily held up owing to a defect in the floats. This defect was found when the aircraft were beached for a minor inspection. The damage seems to have been caused by a fault in design. The design is such that there has been no allowance for water pressure during a turn or during cross-wind take-off and landing. The rear compartment is constructed of too thin gauge material and is liable to collapse around the trailing edge. The makers have brought out a modification which seems to prevent collapse, but will not stop the collapse of the metal in the next compartment forward. This, however, may be prevented by a local modification which has been approved by AHQ East Africa. The local modification, at present under test, consists of strengthening by fitting two stringers; if this does not prove successful, it may be modified to use one stringer instead of two. (259 Squadron ORB, March 1945)

This was followed by a 'bolt from the blue' in early April when 259 Squadron was told to cease all training and to disband. This harsh blow was only slightly lessened when a follow-on instruction stated that aircraft and crews were to be sent to Durban to equip and convert 35 Squadron SAAF.

The only operational sorties by 270 Squadron took place in March and April when a detachment of four aircraft operated from Jui on anti-submarine patrols, although a number of met flights were also flown. All flying ceased for the unit on 15 May, the squadron having been told that it would be returning to the UK. In the event, 270 Squadron became another casualty of the general run-down and was disbanded on 30 June 1945. The New Zealanders of 490 Squadron survived to 1 August before they too were disbanded.

The end of the war with Germany brought an immediate end for 95 Squadron, all operational flying ceasing on 25 May, following which a number of aircraft were flown back to the UK whilst the others were scuttled in the Gambia. Indeed, the 95 Squadron ORB closes with a sad note in June 1945:

> C/95, rocking at anchor at Half Die, received a mortal blow from a native dhow, which would have rendered her unserviceable for some time to come ... in the afternoon of June 26 the last remaining aircraft was taxied slowly from Half Die to the destruction position. Having been fired by the pinnace, C/95 went majestically down to join her sisters beneath the waters of the Gambia. (95 Squadron ORB, June 1945)

A poignant note – and one wonders what condition these sunken relics are now in.

CHAPTER SEVEN

Far East

230 Squadron's L2164 in 1939; the squadron had been operating Short Singapores at Seletar since 1937 and re-equipped with Sunderlands in summer 1938. Peter Green Collection

Flying boats and seaplanes had always played a part in the RAF's strategic plans in the Far East, there being plenty of water and not many airfields. 230 Squadron had been resident at Seletar, Singapore with, appropriately enough, Short Singapores since January 1937. It was a logical move to re-equip the unit with Sunderlands.

The first aircraft destined for 230 Squadron, L2159, left Pembroke Dock on 9 June 1938 and arrived at Seletar on 22 June having routed via Gibraltar, Malta, Alexandria, Lake Habbaniya, Bahrein, Karachi, Gwalior, Calcutta, Dalar River, Rangoon and Mergui. Further aircraft arrived in subsequent weeks, although the Squadron had still only received four aircraft by mid-August.

In October a number of 230 Squadron's Sunderlands had been 'christened' by various regional potentates, the first being L2160, which became 'Selangor' when christened by the Sultan of Selangor at Port Swettenham. It was followed by L2164 'Pahang' and L2161 'Negri-Sembilan'. These three aircraft had been funded by a £300,000 gift from the Federated Malay States Sultans. A Colonial Development Cruise was flown by three aircraft of 230 Squadron in December to:

1. Examine Trincomalee, Colombo and Galle harbours as potential seaplane bases;
2. Enable the AOC (AVM J.T. Babington) to inspect Air Ministry works at Trincomalee;
3. Carry out long distance operational training.

On the 12 December leg of the journey L2161 had a major problem: 'The port outer engine seized and the airscrew sheared, this in turn damaging the port inner so that it too was lost. The aircraft diverted to Nancowry.' The RAF Auxiliary *Aquarius* arrived on 23 December with four spare engines and two airscrews, but little work appears to have been carried out for some days. A week later, on 30 December, Sunderland L2160 arrived with fuel for the stranded aircraft; however, on take-off from Nancowry it suffered a drop in oil pressure on one engine and made a precautionary landing at Great Nicobar Island. In order to prevent the engine seizing, the propeller was lashed fast and the aircraft flew back to Nancowry on three engines where, in a 36-hour period, it was given one of the spare engines brought in by the *Aquarius*, enabling it to return to base on 2 January. A few days later, 5 January, Sunderland L5804 arrived with a spare engine and more fuel, as well as additional engine fitters to speed the work along. A final fuel uplift arrived on 10 January and L2161 carried out a flight test – which proved unsatisfactory. However, after a few adjustments a successful flight test was flown and the following day both Sunderlands flew to Penang, returning to Seletar on 12 January.

Apa Khebar, the magazine of RAF Seletar recorded this incident in its March 1939 issue:

> On December 12, 1938 one of the Short Sunderland flying boats belonging to No. 230 (GR) Squadron and returning from Ceylon to Singapore made a forced landing with two seized engines at Nancowry harbour in the Nicobar islands. She had no spare engine parts on board, so that RAF personnel had to take some out to her, as well as water, food, petrol and other necessities (including beer!). Some of the men spent Christmas and New Year there. One Sunderland was sent from Seletar on December 28 to evacuate an acute appendicitis case, but after an hour's journey out from Nancowry on its return on December 29 it had to turn back owing to engine trouble and make a forced landing of its own approx 50 miles north of Nancowry. It was able to set off again, however, on three engines and reached Nancowry where a spare engine was fitted in two days (a record?); the flying boat returning to Seletar on January 2.
>
> Surrounded by the Indian Ocean, these islands appear to have been meant by nature to be kept aloof from the rest of the world – untouched by social, economic, and political forces. They are administered by a representative of the Indian Government, Dr Sapre. His headquarters were about 3 miles away from the moored Sunderland, and he gave the RAF his closest co-operation and aid during their stay. When the flying boat first landed he confessed that he was somewhat alarmed. Seeing the Jawi characters on the fuselage (the aircraft being one of the four named after the Federated Malay States) he thought it was a Turkish or Egyptian aircraft. Having not heard that the September crisis had passed off peacefully, he thought that we had come to seize the island.
>
> Queen Ishlon, the ruler of Champion Island, which was nearest the flying boat, was also interested in our welfare, and even presented the personnel with a cooked pig on Christmas Day. On the same day she invited our people into her palace, all gaily decorated with images of friends and devils. Her courtiers did a series of war dances, one to the tune of 'Clementine' which Dr Sapre thought must have been introduced into the islands by a visiting methodist missionary years previously.

The article then went into poetic mode:

> It was Xmas Day at Nancowry,
> the crew had been working all day,
> The sun had set in the heavens
> It was calm in Spiteful Bay.
> The Dinner was finished and done with
> The beer had begun to flow
> When up spoke a gallant young officer
> Who called out 'All hands below'.
> 'Man-eating croc. On the Port Bow'.
> The lads all rushed to the deck,
> They stopped and they looked and listened
> And I'm darned if it wasn't, by heck.
> But with the dawn came the reason,
> Sore heads and dry throats too,
> When they stopped and looked and listened
> 'Twas a dirty old native canoe.

The 230 Squadron ORB contains a delightful colonial turn of phrase in a February entry when a Sunderland was tasked on an anti-submarine patrol for a 'submarine of a foreign power believed to be in the vicinity of Singapore without Admiralty sanction'.

Sunderland L5801 crashed on take-off in the Johore Straits on 5 June 1939, two crew being killed. The aircraft subsequently sank whilst being towed; it was later salvaged by the naval base, but then declared a write-off. Interestingly, the replacement aircraft, N9029, which arrived on 25 July, is recorded as having 'the modified bomb aimer's hatch' as well as being camouflaged. Unfortunately, no further details of the hatch are recorded.

The first operational Sunderland sortie in the Far East was flown on 9 September by Flt Lt P. Alington in 230 Squadron's 'Z', the sortie being an anti-submarine sweep ahead of HMT *Neuralia*. However, although the outbreak of war with Germany brought little in the way of operational sorties it did result in a flurry of activity for 230 Squadron.

Problems with Bases

In mid-October OC 230 Squadron, Wg Cdr Bryer, was ordered to deploy aircraft to Malaysia and Ceylon; the squadron ORB details the trials and tribulations of this endeavour. It is worth looking at this in some detail as it demonstrates the problems of establishing operating bases with limited resources, especially when attempting to provide facilities for such a large flying boat as the Sunderland, and in the face of authorities reluctant to accept that war had broken out. On 15 October three aircraft moved, 'on an oral order from AOC Far East', from Seletar to Glugar Airport, Penang, three more making the same journey the following day. All six aircraft were immediately prepared for operations, with 1,650gal of fuel, five days' rations, 25gal of drinking water, and two 250lb SAP bombs. 'However, the senior naval officer claimed no knowledge of the *Admiral Scheer* (our suggested target) or any other enemy ships' (230 Squadron ORB). Indeed, the Navy was surprised at the presence of the RAF aircraft, as were the airport authorities and the fortress commander, none of whom appear to have been informed:

> ... they had received no warning of aircraft arriving and (were) reluctant to accept the CO's explanation that service aircraft may move without notice in consequence of war requirements. Fortress commander also protested on principle, but admitted that he had no guns with which to open fire. Aircraft actually approached below 2,000ft and were not challenged. NOIC [Naval Officer in Command] protested to Commodore Malaya that he had not been informed and was therefore unable to warn Fortress Commander and HM ships.

The following day was no easier, the ORB recording that

> ... the Imperial Airways agent represented that if one of the buoys occupied by a Sunderland was not vacated, he would stop Imperial Airways aircraft at Singapore and Bangkok, thus

> interrupting the service. He pressed that we occupy another buoy, which he admitted Imperial Airways pilots regarded as unsafe. CO decided that HM aircraft on war service, heavily loaded and carrying explosives, should have preference and reported to HQ Far East.

This reluctance to accept that war had arrived was to plague the squadron, and other British military organizations in the Far East, for some months; indeed, the eventual fall of Hong Kong and Singapore is in many respects a direct result of this attitude amongst civil authorities and complacent military commanders. 18 October brought a reply from HQFE

> … instructing the CO to anchor one Sunderland to allow Imperial Airways to use mooring. CO, as senior officer on the spot, decided that this was an unwarranted risk, as Imperial Airways Flying Boat was much lighter and could go to Singapore after a short stop. Examined Harbour Master's extra mooring, the one refused by Imperial Airways, and concluded that it was safe in fair weather; aircraft towed to it with outboard engine in a sampan, pilots kept on board.

The following day the support ship *Tung Tung* arrived and laid the three moorings that it carried, thus relieving the situation.

Operation Instruction 503 was issued on 22 October to cover future possible moves. The AOC having implied that such a move should be kept secret, the four likely destinations were given code-names: Trincomalee (ACE), Nicobar Islands (QUEEN), Seletar (DUCE), and any 'unanticipated destination' (KING). The same instruction enforced W/T silence en route to Penang and whilst on the water at Penang.

A move to China Bay was ordered on 26 October, six aircraft leaving Penang at 0600 the following day. Flt Lt Smith's aircraft developed

> … big end failure in starboard inner engine and alighted [on the water]. Sqn Ldr Francis alighted alongside and gave instructions. Airscrew secured to exhaust ring, petrol jettisoned down to 350gal, bombs jettisoned, spare oil, dope and sundries jettisoned. Took off, with some difficulty, owing to swell and tendency to swerve, and flew to Nancowry, approx. 200 miles away. (230 Squadron ORB)

The situation at China Bay was to prove even more difficult than that at Penang and although OC China Bay had improvised a landing jetty and raft, and had laid six moorings in Malay Cove, the 'government slipway' was too short and too weak for a Sunderland. Likewise, the initial squadron assessment was that Trincomalee was 'unprepared and unpromising'. On 28 October Wg Cdr Bryer had a meeting with the C-in-C, Vice-Admiral Leatham, who expressed his surprise that he did not know the RAF were coming, but hoped that they would mount regular patrols. The CO agreed to a patrol routine, one over a 'focal area about Colombo and the other about the Maldives'. The same day Coastwatch reported a submarine and two Sunderlands were sent to investigate, with no result; one aircraft damaged a float on landing at Trincomalee. As the RAF seaplane tender proved useless, the CO arranged hire of a large paddling canoe and a private motor launch in order to effect repairs.

RN294 of 230 Squadron over the jungles of Ceylon; a Blackburn-built Mk V, the aircraft subsequently sank at Wig Bay in December 1951. Peter Green Collection

Koggala had been home to 230 Squadron for much of the war. Ken Delve Collection

With the likely requirement for additional operational locations, two aircraft flew to Koggala Lagoon on 30 October to take soundings and assess the lake's suitability for Sunderland operations. Two aircraft also went to Colombo:

> None of the promised naval and military co-operation materialized. The Shell Company provided a boat, arranged petrol supplies and generally helped in lieu of the Navy. The officers were accommodated in the Grand Oriental Hotel and the airmen in the British Soldiers and Sailors Institute.

Flt Lt Garside was sent on 1 November to

> ... reconnoitre local resources at Galle and Koggala. Barges can be taken in by sea route and small country boats are available, but no pulling or motor craft suitable for communications between aircraft and shore. No accommodation near Koggala, a few houses, none easily adapted, in Galle. Motor buses between Galle and Koggala not suitable for Europeans; trains infrequent but possible for restricted routine and for goods to ½ mile from the water.

The ORB is still not very complementary about the 'main base' at Colombo:

> There are no motor boats in Colombo harbour suitable for attendance on flying boats. Pulling boats were used to ply between flying boats and a hired launch. Shell Company provided a barge with petrol in drums. Officers scoured all likely places near Colombo for suitable boats. A private boat was found 12 miles away, brought in by canal and submitted to Admiralty Overseer, passed and taken into service. Slipway arrangements examined: 1. Government Patent Slip will necessitate manoeuvring in area subject to passage of strings of barges, under poor control, and fine clearances between floats and wing-tips and obstructions on shore; 2. it would be possible to make a slipway at sill on north side of guide pier, no other alternative recommended in Colombo, and both 1. and 2. would be subject to adverse sea and wind conditions in SW monsoon and occasionally in NE monsoon. Alternatives worthy of consideration elsewhere are: 3. strengthening slipway at China Bay, but probably need extension as well, not recommended owing to cross-winds, rough water, rocky bottom off slipway, obstruction to hangar, and apron too small; 4. making slipway at Naval Boom Defence Depot – advantages, deep water just off existing large apron used for spreading nets, winch, railway, workshops available, facing NE monsoon, sheltered from SW monsoon; disadvantages, interference with boom defence, not on Air Ministry property; 5. new slipway planned for China Bay – early construction unlikely and policy likely to be revised; 6. Koggala Lake, 10 miles east of Galle – calm water all year round, material available locally but no accurate survey. Note no other locality in Ceylon

Sunderland III ML865 cruises along the coast of Ceylon. Arthur Banks appears to offer a site except by extensive and costly dredging. Alternative 1. to be tried. (230 Squadron ORB, November 1939)

By 3 November the Squadron had proposed to have

... four Sunderlands ready to operate from Colombo as this is nearer the focal areas of shipping than China Bay; intelligence and operations room facilities available (including only reliable communications) in Naval office, personal liaison with C-in-C and Naval Staff. Disadvantages – no discipline among harbour craft when flying in progress, night flying not practicable except in bright moonlight.

The following day it was decided to establish a mooring at Male/Sultan's Island, Maldives, along with a small stock of petrol and oil. At Colombo, congestion of the harbour and the possibility of damage to the Sunderlands led the CO to suggest an operational move to Koggala, with personnel accommodated at Galle. The Shell Company was ordered to send 2,600gal of DTD 230 petrol to Galle as part of this move, whilst the Surveyor General was tasked with surveying the shore at Koggala with a view to making a flight path from the lake for use in the SW monsoon, and providing areas for working, handling petrol, and accommodation for guards.

As if these problems were not enough to cope with, the Squadron was having difficulties with spares:

[limited] flying has been permitted owing to the acute shortage of exhaust manifolds and tail pipes, and a probable shortage of Pegasus XXII engines. Informed today [10 November] that a few urgently required stores despatched in HMS *Eagle* from Aircraft Depot, Far East. Intensive operations would have brought Squadron to a standstill in about 150 hours' total flying.

The only positive note recorded in the ORB related to the excellent co-operation being given by the Maldivian Government, their representative having promised all possible help because 'we have enjoyed 150 years of unbroken peace and prosperity, without interference with our customs and internal affairs, under the protection of the British Empire'. The point was well taken – and no doubt also aimed, through the ORB, at other, less co-operative, agencies.

A signal was sent to AOC India informing him that the squadron was conducting a reconnaissance of Cochin in respect of the possible use of this location for emergencies; meanwhile, C-in-C East Indies requesting an examination of the prospects of stationing flying boats at Port Blair, Andaman Islands, off Nancowry.

It just didn't seem to get any better and the ORB recorded on 25 November that the CO was told that 'a RS250 fine would be imposed if work was carried out on aircraft on Sundays. Government departments do not recognize that peace-time regulations must be relaxed in war-time.' The squadron now had four aircraft at Koggala, with 1,800gal each, one at China Bay, with 1,600gal, and one under repair at

240 Squadron's 'M' rests on its mooring in Penang. Ken Delve Collection

Colombo. This was quite reasonable serviceability for the Sunderlands at the time, but a warning was sent to HQFE that 'exhaust manifolds were disintegrating whilst the aircraft sat at their moorings, especially those that had been repaired by welding. New ones are urgently needed.' The supply situation had already been poor before the squadron had left Seletar: by the following January the squadron was reduced to trying out improvised repairs: 'Sunderland 'Z' was flown with a temporary repair of plasticene, pads and struts to see if the repair would remain intact – it stripped off during take-off.' Spare parts were not the only operational elements in short supply: the CO pointed out that his crew situation was not good in respect of experience, with only six qualified First Pilots, four part-trained Second Pilots and two part-trained Sergeant Observers, along with two pilots on extended sick leave.

Such a detailed 'warts and all' record is, sadly, all too rare in Squadron official records; the 230 Squadron compiler and CO have provided an excellent insight into what was happening at the time.

December brought authority from HQFE for 230 Squadron to operate at an overload weight of 56,000lb (25,300kg), although this was restricted to 52,000lb (23,600kg) for operations from Koggala.

As 230 Squadron was now fully established at Koggala, Ceylon, with effect from 13 February, the remaining personnel at Seletar were designated 'S Flight' of 205 Squadron (soon to receive Catalinas). A few months later the squadron bade farewell to the Far East, en route for Middle East Command. Sunderlands would not return to the Far East until 1944.

Back in Theatre

Although the Japanese had opened a submarine base at Penang, Malaysia, in 1942 it was only the following year that this was used by German submarines as well, the latter partly involved in anti-shipping work but also, increasingly, in blockade running. By the latter part of the year intelligence assessments suggested that some fourteen U-boats, seven Japanese and seven German, were operating in the Indian Ocean. Shipping losses had risen steadily throughout the year and in view of the Allied military build-up and plans for early 1944 it was decided to strengthen both the organization and assets of the anti-submarine forces. The air commander had ten flying boat squadrons, only one of which was equipped with Sunderlands, in Ceylon, India and East Africa, operational control being exercised by No. 222 Group.

In February 1944, 230 Squadron made the move to Koggala, Ceylon, the squadron recording that

> Twelve far from unpleasant months had been spent in East Africa. Eight aircraft left Dar es Salaam and proceeded via Kisumu, Khartoum, Aden, Masirah, Korangi Creek, Bombay and Cochin. A ninth aircraft followed in March.

The squadron was to operate alongside two Catalina squadrons, Nos 205 and 413, as part of No. 222 Group. The first operation in their new theatre was flown on 1 March by Flt Lt Watson in EJ141, a U-boat hunt east of the Chagos archipelago, no sightings being made in the 13-hour sortie. Sunderlands flew similar missions over the next two days but no submarines were spotted. However, the first ASR mission, 5 March was more successful, a Sunderland locating the survivors of a merchant ship that had been torpedoed. The sea was too rough for the flying-boat to land but an airborne watch was kept until rescue arrived.

An intelligence assessment in April concluded that there were only two U-boats active in the Indian Ocean; however, the level of anti-submarine patrols for all three flying-boat squadrons was maintained. The following month, 230 Squadron was tasked on Operation *River*, the evacuation of battle casualties and sick personnel from the campaign in north Burma.

The spring 1944 Allied advance into Burma, aimed at capturing Rangoon, relied heavily on air support. The Sunderlands of 230 Squadron played a small but significant part in these operations. One of 14th Army's main logistic systems involved the Inland Water Transport (IWT) Service using the Chindwin and Irrawaddy rivers. For a number of weeks in late February/early March the Sunderlands flew vital sorties carrying heavy marine engines for the IWT from Bombay, via Bally (Calcutta) to a point on the Chindwin near Kalewa. Long-awaited exhaust spares arrived on 21 March from China Bay and aircraft were serviced. A few days later news was received of two enemy warships in the Indian Ocean and all aircraft were brought to 24-hour notice for a move to the Seychelles.

The RAF Auxiliary *Tung Song* arrived on 2 April to lay an additional five moorings and deposit 6,400gal of DTD 230 aviation fuel, the ship sailing for the Seychelles where it left a further 25,000gal.

General Orde Wingate had initiated his second Chindit campaign into Burma in early March, Operation *Thursday* landing troops at three locations (Broadway, Piccadilly and Chowringhee). Casualties began to mount and the decision was taken to evacuate the wounded as quickly as possible, usually with a 'supplies in'/'wounded out' system. Lake Indawgyi was chosen for operations by Sunderlands, the marshy nature of the area precluding the use of large, land-based aircraft. The first successful landing on the lake took place on 2 June and a second aircraft arrived at Dibrugarh a few days later. Two aircraft, nicknamed 'Gert' and 'Daisy' by the Army, operated from Dibrugarh on the river Brahmaputra on this evacuation work from 31 May, some 500 casualties of 3rd Indian Division being taken out during June and vital supplies taken in. One of the aircraft, 'Q', was wrecked at Dibrugarh on 4 July by a hurricane.

Although few U-boats were active they were taking a toll: in July five ships (100,000 tons) were sunk. This was, however, to be the last month with such losses and U-boat activity subsequently declined.

In August the 230 Squadron records note a series of drogue and air-to-sea firing exercises 'in connection with tests carried out with 0.5in Brownings from the Bomb Aimer's position'. Unfortunately, no further details are given.

Sqn Ldr K. Ingham (JM673/P) failed to return from a sortie on 28 November; over the next few days the squadron flew a large number of sorties searching for this popular Flight Commander and his crew, but to no avail.

The major change for 1945 was the arrival of the Sunderland V in theatre, 230 Squadron's ORB recording 'The Squadron is due to re-equip completely with twelve UE Sunderland Mk 5 aircraft, fitted with four Pratt & Whitney 1830-90B series engines and Mk III ASV. The present establishment is nine UE.'

January saw the first aircraft arrive on 230 Squadron, leading to a reduction in operational flying and an increase in training: 'All aircraft to leave the Squadron had to be prepared for the 50-hour flight to the UK. A series of lectures on working and maintaining of the new aircraft were programmed and all Captains flew circuits in the new aircraft.' The first Sunderland V had arrived, via Korangi Creek, on 6 January, flown by Flt Lt Potter:

> At 31st January only four Mk V Sunderlands had been delivered ex UK [the original plan was for eleven to have been delivered by this date]. The aircraft are delivered at Korangi Creek (374 MU) by Ferry Crews from the UK and the Squadron flies in a Mk III for storage purposes and collects a Mk V for delivery to Koggala. It is considered that an excessive delay occurs between the time aircraft are delivered to Korangi Creek and the time they arrive in the Squadron, due mainly to the fact that the Squadron is not informed of their arrival at Korangi Creek. In the case of one Mk V the aircraft was kept at Korangi Creek for eighteen days to enable it to be used in an experimental capacity for trying out beaching chassis and the use of the slipway generally. The morale of the Squadron was raised considerably due to the arrival of the Sunderland Mk V. The enthusiasm of the groundcrew has been somewhat dampened by reason of the fact that the long awaited 'pack up' had not arrived by the target date of January 27. In particular, ground-handling equipment is non-existent, or in short supply, beaching chassis, engine stands and Daily Inspection and Maintenance Stands are the chief items awaited. (230 Squadron ORB)

The record continued:

> The arrival of the first Sunderland Mk 5 caused a great deal of interest amongst all personnel at Koggala and it appeared as though the entire station had turned out to watch its arrival. Flt Lt Potter gave the onlookers an excellent opportunity to see for themselves the superior performance of the Mk 5 as compared to the Mk III, when he put the aircraft through its paces. There can be no doubt that all were impressed by the exhibition especially the climbing performance of the new aircraft, and the Squadron felt justly proud that 230 should have been selected for the honour of being the first squadron to be re-equipped with Mk 5s.

A second aircraft, PP146, arrived in the afternoon.

Having been operating as a Catalina unit since April 1941, 209 Squadron at Kipevu was scheduled to re-equip in early 1945 with Sunderlands, the first aircraft, Mk V PP151, arriving from Kisumu on 21 February. Two days later Flt Lt Fumerton was attached to the squadron to instruct pilots in Sunderland flying and procedures.

The first Sunderland V to arrive in East Africa for 209 Squadron, February 1945. Arthur Banks

(Right) **The nose Brownings are just visible on this Sunderland V of 230 Squadron.** Peter Green Collection

PP151 was officially handed over the same day, the 'aircraft being hauled up the slip in the afternoon, various difficulties being overcome in the process. A minor inspection commenced the following day, various aircrew personnel attended and helped with this inspection so as to get as much "gen" as possible' (209 Squadron ORB). Having had its compasses swung, the aircraft was launched and moored by the end of the month, and then spent most of March in use for conversion training, being joined by more aircraft as the month progressed. On 28 March the CO, Wg Cdr Woodward, held a squadron parade and announced that 209 Squadron was 'scheduled to be fully operational as a night attack squadron by the end of April and stressed the fact that all personnel must be prepared to put in extra time on ground work and training'. All operational flying was being conducted by the remaining Catalinas, the last of which departed in May.

A change of role was also promulgated for 230 Squadron:

The month of March saw the beginning of a new role for the Squadron. It was not without much speculation that the Squadron accepted the more reliable and improved Mark V aircraft. Suddenly new shipping exercises were laid on with three aircraft taking part in one sortie. It was learnt at the briefing that our role was to attack shipping, three aircraft to first of all carry out a parallel track search, locate the target and 'home' the remainder. Then circling the target at 120 degree intervals and outside the estimated range of flak, to make a concerted attack on the target, one aircraft being selected to do the bombing by a complete run over the target, the remainder to act as anti-flak to the bombing aircraft, and breaking away at 400yd. In view of our new role considerable training was done in air-to-air firing, fighter affiliation with 81 Squadron, and bombing exercises. The next thing to be learnt was that the Squadron was to attack Japanese coastal shipping off Burma and the CO, at a 100 per cent Squadron parade, informed all personnel that they would in future operate from a forward base on the depot ship *Manela* and that all would have to forget the pleasant life of Koggala to the cramped conditions of ship life. All sections packed up 50 per cent of their equipment and transported it to Colombo to put on the *Manela*. (230 Squadron ORB, March 1945)

Further details were provided in the following month's record:

The *Manela* was based at Akyab the intention being to convert the Squadron to a more mobile role so that it can operate as closely as possible with the advancing forces in Burma and Malaya. On arrival at Akyab arrangements had not been made for moorings, so that alternative ones had to be found in the Roadstead 4 or 5 miles away from the proposed alighting and mooring area. The CO, Wg Cdr C. E. L. Powell, set up the Squadron HQ on April 17 and the rest of the advance party had arrived by the 20th.

By late March the squadron had received its planned establishment of twelve Sunderland Vs and 'the groundcrew worked hard to ensure that all the Squadron's aircraft are in first-class condition for the forthcoming Squadron move and employment on anti-shipping operations'. The advance party left for Akyab, Burma on 16 April and within days the bulk of the squadron had left Koggala, the latter in future being a detachment under Sqn Ldr Deller. First May saw Fg Off J. Toller in ML799/W fly the first sortie as part of the combined operations covering the landings at Rangoon, a crossover patrol to cover the invasion fleets. The crew spotted two single-seat fighters, which they believed to be 'Sonias', these making a curve of pursuit to within 800yd but which they broke off when the Sunderland entered cloud.

Meanwhile, for 209 Squadron the training routine continued and by April the squadron was flying over 150 Sunderland sorties a month, albeit still no operations. Tragedy struck on 14 May; Fg Off Gibson ('V') was airborne for radar homing practice but was forced to return when the equipment went unserviceable and the weather deteriorated. The weather at base was also poor and he was ordered to circle outside the harbour until it cleared; it was ascertained the next day that the aircraft had crashed in the Quali District and caught fire, three of the crew being killed.

Sunderlands were now flying GR/armed recce sorties on a regular basis, usually with little to report; however, 230 Squadron picked up the Japanese 10,000 ton tanker *Toho Maru* under escort in the Gulf of Siam. This plum target was reported back to base and a strike was mounted on 15 June, one section of Liberators from 159 Squadron managing to score several hits on the tanker, leaving it a blazing wreck.

May saw 230 Squadron transfer its HQ to Rangoon whilst the operational detachments were at Redhills Lake and Koggala, the move being made because Akyab was too far away from the operational theatre. The following month the squadron was flying armed reconnaissance over the Andeman Sea and the Gulf of Siam. Attacks were made on coastal shipping, six such ships being sunk this month.

The squadron also recorded that June

... has proved one of the most trying months in the Squadron's history. There was a constant stream of conflicting orders, originating from departments comfortably ensconced in far-away Colombo, who appeared to have had little idea of the difficulties the Squadron were working under, and still less idea of where they wanted it to be based. At the end of the month we were told to establish base at Koggala and keep detachments of six aircraft at Rangoon or as near the scene of operations as possible, under No. 346 Wing. The recall of all Dominion personnel except the New Zealanders has cut the aircrew by one third and all crews were affected. (230 Squadron ORB, June 1945)

A spot of window cleaning by the aircraft's captain, Les George, at Pamanzi, May 1945. Arthur Banks

The Pegus River at Rangoon was by no means ideal for flying boat operations and 230 Squadron had six Sunderlands with serious hull damage caused by 'contrary winds and the swift running tides in the monsoon-swollen river; four of these accidents being caused by marine craft which were almost uncontrollable in these tides'. (230 Squadron ORB)

By June the squadron was operating various small detachments, locations including Pamanzi, Diego Suarez and Dar es Salaam. Warning orders were received the same month for a squadron move, all aircrew were sent on leave and the opportunity was taken to perform a modification to the blowers on the engines. The first Sunderland duly moved to Koggala, Ceylon, on 12 July, via the Seychelles and Diego Garcia. However, within days a detachment was sent to Rangoon to work with No. 346 Wing. Sunderland PP159/Q was first away, on 21 July, the aircraft carrying four groundcrew (fitter, instrument repairer, rigger and wireless mechanic) in addition to its normal crew:

> On arrival, all personnel were billeted in buildings with No. 346 Wing. The buildings were pretty thoroughly stripped of all furnishings and had no beds until later aircraft brought a few, or until rough ones were made on the spot. … The 7–8kt flow on the river made refuelling and servicing of aircraft difficult, also the fact that one refueller was none too quick at pumping in the gas did not help matters. (209 Squadron ORB)

The detachment soon grew to four aircraft, Sqn Ldr Walker taking command; in the meantime, the first operational sortie had been flown on 27 July, Fg Off Yeomans taking PP159 on an armed reconnaissance off the Kra Isthmus against 'Sugar Dogs and other small coastal craft being used by the Japanese. Four 250lb DCs were carried. One large twin-masted junk was damaged and machine-gun hits obtained on one 100ft twin-masted schooner, one small schooner, two sampans and one raft.' A similar sortie was flown by Flt Lt McKendrick the following day, although this time the main weapons were 250lb GP bombs; four of these were dropped on one ship, but all failed to explode.

The trots at Koggala; 209 Squadron moved to this location in July 1945. Arthur Banks

(Below) **Koggala 1945 and a Sunderland V of 205 Squadron; the squadron gave up its Catalinas in favour of Sunderlands in June 1945.** Peter Green Collection

Mark V PP126 of 240 Squadron at rest in Korangi Creek in July 1945. Ken Delve Collection

Sunderland numbers in the Far East were strengthened in the early summer. Having operated very successfully as a Catalina squadron for some time, the May news of

> ... the disbandment of the old Squadron made a very gloomy end to the month. It is hoped that the new squadron, which has taken over the 240 Squadron identity, will perpetuate the good work which has been carried out since the formation of the old Squadron. (240 Squadron ORB, May 1945)

The changeover was made on 1 July by the amalgamation at Redhills Lake of Nos 212 and 240 Squadrons. One of the Flight Commanders, Sqn Ldr Hoare, and a number of other aircrew undertook a Sunderland conversion course at Koggala, the intention being that they would then convert the rest of the squadron. The first Sunderland V, PP126/K, arrived in early July. Re-equipment was not complete until the end of the year, the squadron's Special Duties Catalina flight continuing its secret missions. At the end of July, 230 Squadron moved from Rangoon to Redhills Lake to undergo intensive training for a new operational role. Indeed, both of the Sunderland squadrons at Redhills Lake were now to take on new roles.

The final operational sorties by Sunderlands in this theatre involved the repatriation of Prisoners of War, the condition of these individuals causing consternation amongst the flying boat crews.

The end of the war brought an instruction to 240 Squadron to remove all offensive and defensive armament, and all guns were removed with the exception of the 0.5in beam guns, the squadron electing to retain these against the possibility of 'resisting pockets of the enemy'. Gun turrets were either removed or blanked in. The squadron now primarily operated as an ASR unit, although routine patrols and Met flights were also flown, plus the occasional sortie supporting 230 Squadron's daily run to Singapore. One ASR sortie involved the location and rescue of a 355 Squadron Liberator crew on 17 September:

Muscle power as PP130 comes up the slip; this aircraft ended its career with 235 OCU before being struck off charge in June 1955. Ken Delve Collection

Fg Off Robinson (NJ273/D) left Rangoon at 2322 GMT. At 0254 saw red Verey cartridge from port waist position and located Liberator at 0257; five survivors were seen on the starboard wing. Flew low over the wreck to check condition of survivors and take pictures. At 0315 dropped Bircham Barrel and message in tin – 'please indicate number of crew injuries'. The reply came back that one man had an injured arm. Contacted base to request permission to land in Sittang Estuary as the local area was too marshy and flew over to check the landing area. Back at the Liberator found two circling Beaufighters. At 0415 base asked if possible to land at Sittang, at 0436 message 'use discretion – taxy taking soundings – uncharted waters'. Dropped another message – 'can you make coast on foot – I will indicate direction and land in sea and pick you up'. The Liberator crew indicated OK and aircraft led them to the coast, natives acting as guides and bearers. At 0445 sent – 'am leading crew to coast, will land'. Aircraft landed safely and taxied to coast, firing Verey cartridges, anchored and sent a party of four ashore in rubber dinghy. All picked up OK, dried and given hot tea. At 0855 taxied to open sea but too rough for take-off, went back to the river mouth to take-off down-river in direction of sea, several times the aircraft ran aground on soft mud but was safely airborne at 0915. Land base 0944. (240 Squadron ORB, September 1944)

Rescuing the crew of a 355 Squadron Liberator on 17 September 1945. The Sunderland landed off-shore and picked up the crew. Ken Delve Collection

It had been a similar situation for 230 Squadron, all non-essential equipment being removed in order to increase the useful payload for the new transport role. 'A close study was made of the seating capacity and the ditching drill for passengers.' Accommodation was arranged for twenty seated passengers along with bunks for twelve stretcher cases. Other changes included the fitting of the Loran navigation aid to eight of the aircraft, the rest being scheduled for equipping when sufficient sets had arrived, along with a plan to fit VHF radios in all aircraft. The arrival of three more aircraft in September brought squadron strength to twelve Sunderland Vs. One of the main tasks was to be a daily courier service to Singapore and the first aircraft flew this route on 10 September, the plan being to fly medical supplies and personnel in to Singapore and to fly ex-Prisoners of War out. By the end of October the squadron had flown out 315 PoWs; for most of the month a detachment of aircraft was kept at Seletar. Sadly, there was one tragedy during this period, Sunderland 'X' (Flt Lt Levty-Haarscher), crashing into the jungle on 15 October with the loss of all the crew and fourteen PoWs.

The Singapore freight/passenger/PoWs routine also kept 240 Squadron busy during October and November, with late October also bringing the Rangoon detachment to an end, although the detachment at Penang was maintained. Indeed, the latter was to stay in place into 1946, the main element of the squadron having moved to Koggala in mid-January. This move reflected the needs of the courier service from Koggala to Seletar that had been operational since December.

The flying boat's role in such tasking was singled out for comment:

On a few outstanding occasions, these [Sunderland and Catalina] squadrons blazed the trail which others were to follow. They occurred in September 1945. ... On 10 Sept 45 of the first six aircraft to reach Singapore, five were Sunderlands (No. 209 Squadron). They established the flying base at Seletar; on 13 and 14 Sept respectively vanguard Sunderlands of 205 and 230 Squadrons followed. On 18 Sept, the first Sunderland was waterborne at Hainan Island. On 28 and 29 Sept the first two Sunderlands (both of 209 Squadron) were waterborne at Tandjoeng Prick anchorage at Batavia and evacuate APWI to Singapore. On 1 Oct the first aircraft (Sunderland of 209 Squadron) flew to Kuching, China and, soon after, a detachment of Sunderlands was anchored at Kai Tak (Hong Kong). On 2 Dec the first Sunderland flew to Labuan in Borneo and reported excellent anchorage for flying boats. About the same time, the first Sunderland reached Shanghai and, during that month, the flying boat anchorage at Catlai (7 miles east of Saigon) was brought into use. (AHB Narrative, 'The RAF in the maritime war')

In August No. 205 Squadron despatched four Sunderlands to the Cocos Islands and to Singapore laden with RAPWI and Red Cross stores. They made about sixteen flights back from Seletar to Koggala with varying numbers up to fifteen RAPWI personnel. No. 209 Squadron's effort ranged as far as Hainan, Hong Kong, Saigon and Penang. Like their sister

On the slip at the Seychelles. Arthur Banks

209 Squadron aircraft being beached at the Seychelles after losing a float. Arthur Banks

squadrons at Koggala, No. 230 Squadron at Redhills Lake carried in huge quantities of supplies to Singapore. No. 240 Squadron spent September initiating ferry runs to Bangkok and Hong Kong.

In the month of October, notorious for its bad weather, the four squadrons toiled unceasingly to link up and supply the needy areas and distribution bases. No. 209 worked for the most part on the ferry, or so-called courier, runs between Ceylon, Singapore, Hong Kong and often back to Rangoon. No. 230 Squadron carried stores into Singapore and evacuated RAPWI, while No. 240 Squadron ferried between Madras, Singapore, Bangkok and Saigon. November saw the Sunderland squadrons reverting gradually to the tasks of supply and normal passenger freight, but they still co-operated to a valuable extent to the later phases of Operation *Mastiff*. (AHB narrative)

The month of October came to a close with the main task of repatriating Allied ex PoWs practically completed. The Squadron's effort has accounted for 32,978lb of freight and 196 passengers – all these between Seletar, Penang and Redhills Lake. Weather conditions in general were bad and on many occasions sorties were postponed for two or three days due to a deep depression forming in the Bay of Bengal. At Redhills Lake the arrival of the NE monsoon, with its low cloud and heavy rain, necessitated take-offs being deferred until first light, which in turn meant staging through Penang.

The Squadron detachment at Seletar was called upon to perform varied tasks, the most interesting of which was to transport supplies of rice to the State of Pehang, whose population was faced with imminent famine. It was not possible for surface ships to assist owing to the danger of unswept minefields which lay off Deccan and Kuantan. Sqn Ldr Nicholson was captain of the first aircraft to make the flight to Kuantan, where a safe alighting was made in the river, with a cargo of 10,000lb of rice. There was a considerable amount of risk attached to these missions as there were no charts of the alighting area nor any means of obtaining reliable information as to the depth of water, obstructions, sandbanks, etc. The alighting area was approx. 100yd wide and one mile long.

Flt Lt Moonlight discovered these problems on 3 October, his second sortie into Kuantan, when his Sunderland struck a submerged object, ripping out the bottom of the hull; he beached the aircraft on the nearest sandbank but it was a write-off. After this incident the payload for missions was reduced to 6,000lb. By the time the commitment ended eleven sorties had been flown, carrying 90,000lb of rice.

November saw 230 Squadron move from Madras to Singapore, the commitment to fly PoWs from Malaya having been terminated earlier in the month. On each sortie into Singapore, 1,000lb of freight was allocated to the squadron to allow it to move the majority of its servicing equipment before the main party arrived.

December brought another evacuation task, the Sunderlands flying refugees out of Batavia. One of the immediate problems faced in the Far East after the war ended was the hostility towards European attempts to re-impose colonial control of certain territories, 'freedom fighting guerillas' armed by the Allies to fight the Japanese now seeking independence. The Netherlands East Indies was one particular flashpoint in late 1945 and early 1946.

The Batavia task ended in early January but in the same month the squadron was called on to operate a detachment from Labuan to support Allied forces in North Borneo, including providing reconnaissance of the Dani and Kateman Islands following reports of native uprisings, while also providing ASR support for Spitfires engaged on similar tasks. The living conditions at Labuan were rough: 'accommodated in tents on the beach, food rationed and occasional shortages meant flying with no breakfast. Coral snakes, sharks, crocodiles, black fly, all added to the discomfort.' Additional landing areas were checked in Borneo and the Celebes, Balikpapen, Sandakan and Macassar all being considered suitable. However, the post-war demobilization of personnel was having an increasingly adverse effect on manning, both amongst aircrew and groundcrew.

209 'Z' being pulled up the slip at Seychelles after losing a float. Arthur Banks

Flt Lt Halliday was tasked to take an ENSA party to Hong Kong on 9 February but this simple trip turned out to be quite dramatic. Having cleared an engine problem after take-off the Sunderland set course for Hong Kong; it later became apparent that with stronger than expected headwinds the aircraft would not have enough fuel to make Hong Kong and so the Captain elected to divert to Saigon. He had the unenviable task of alighting in an unknown area, but to lessen the risk followed the wake of a small coaster into a safe landing near Port de Feu Rouge. There was no fuel available at the French barracks so the aircraft took off again and made for the French Naval Air Station at Cat Lai further up river – but there was still no fuel available. The second pilot, Flt Sgt Millington, commandeered a British naval launch and took the ENSA party to Saigon where they were accommodated overnight whilst fuel was obtained from the main airfield at Saigon. The refuelling was accomplished by positioning a steam tug under the wing and using Mae Wests as fenders. After three attempts the Sunderland managed to take off and landed at Hong Kong 7 hours later.

Re-organization of the RAF's post-war requirements caught up with the Far East in spring 1946, one of the first casualties being 240 Squadron, the unit disbanding on 31 March at Koggala. For 230 Squadron there was an increase in establishment to twelve aircraft, but also record levels of unserviceability. Other changes to the aircraft included a trial installation of a TR1173 VHF radio. The squadron ceased all flying, other than the weekly service to Borneo, in late March in preparation for its move to the UK, this move taking place in mid-April.

On 1 September, No. 1430 Flight, at Kai Tak, Hong Kong, was renumbered as 88 Squadron, under Sqn Ldr Heinbuch and with a complement of six aircraft and six crews, its primary task being 'bi-weekly transport service Kai Tak, Hong Kong and Iwakuni'. The first sortie by the 'new' squadron was made the same day, Flt Lt Thomas taking two passengers and 2,950lb of mail/freight to Iwakuni whilst a second aircraft flew from Iwakuni to Kai Tak with 'twenty-one Japanese war criminals whose good behaviour was assured by four Gurkha guards who stowed their rifles on entering the aircraft and brandished kukris in close proximity to the unwelcome passengers'. (88 Squadron ORB, September 1946) In a typical month the squadron would carry about 150 passengers plus 17,000lb of mail and 30,000lb of other freight. In November one aircraft, NJ272/A, was given VIP seating in the Officer's Ward Room, the first passenger to take advantage of this new 'luxury' being Lord Tedder, Chief of the Air Staff.

Gerry Paine recalls one early incident on the squadron:

On September 7 Hong Kong had one of its periodic typhoon warnings and most of 88 Squadron's aircraft flew off to Seletar. Our aircraft, PP130/N was up the 'slip' at the time so it was quickly launched and re-fuelled. The wind was increasing considerably by this time and the sea was quite rough. Standing on the wing helping to refuel, I watched a Dakota taking off in the direction of the mountains ; just as it was clearing the mountains it stalled and crashed in a ball of fire. Soon after, we had a very bumpy take-off as the sea was by then very rough. Our flight took 6½ hours to the US Navy base at Okinawa. On landing we could not easily moor up to their type of buoys, so it being a very hot day and glad of a chance to cool off, I jumped over the bows – only later did I hear that the waters were shark-infested! The next day we flew back to Hong Kong; about two hours from base I was sitting in the Flight Engineer's position when I smelt burning. It rapidly got worse and flames appeared just inside the port wing root. By this time other members of the crew were on hand with fire extinguishers and the fire was quickly put out. The fire had been caused by a 'short' in the Radar Unit.

Routine servicing of the Sunderland was anything but. There was a marine craft section for starters; the Air Force's own sailors, with their floating fuel bowsers, ammunition scows, lighters, pinnaces, fire tenders and crew transport. I'd ride a launch out to the aircraft, bobbing at its mooring in the Straits of Johore, climb aboard and complete my task. But this aircraft was certainly an advancement on earlier types of flying boat, partly due to the wing leading edge alongside the engines being hinged. These hinged sections could be lowered to make a platform on which to stand when working on the engines. But, due to the location and immediate environment, tools needed to be secured were they not to be lost forever. Oops! … clink … plop … damn and blast! Look out, fish, here she comes: $\frac{5}{16} \times \frac{1}{8}$ Whitworth o/e. Inventory holders were never easy to find!

Servicing complete, I'd reach for the Aldis Lamp and signal the piermaster, requesting the launch's return, which could take time. So, imagine, the intrepid fitter, 90+ in the shade, palm trees on the nearby beach barely stirring in a lazy breeze, sea as flat and smooth as unrolled silk, a wing that made a perfect diving platform … but paradise did have the odd flaw: my swimming hole was full of venomous snakes. Like most snakes they offered no harm

PP132 was one of the aircraft 'damaged beyond repair' by a storm at Kai Tak on 21 April 1946. Ken Delve Collection

unless threatened, but trying telling that to a naive airman. Still, I never heard of anyone being attacked when swimming or beaching the aircraft; it seemed the main danger of these impromptu swimming sessions lay in getting back aboard the aircraft.

So excited was I that only brief fragments of my first waterborne take-off are lodged in my memory. A launch had first made a sweep down the 'runway' checking for debris, our engines were running, we'd cast-off from the buoy, were lined up and ready, then with throttles pushed open and the roar of four engines at full power drumming my ears, the spray flew back, obscuring the view – but we were on our way. There was a reluctant build-up of speed – degree of reluctance dependent on hull cleanliness (barnacles being the problem) – as hull and wing-tip floats cut creamy furrows in turquoise water. 'On the step', and the foam subsided as we charged across the surface. A lunge, a couple of bounces, and we were airborne, water-streaked perspex clearing as we clawed our way into the sky. A boat had become an aeroplane. Fleeting but emotional recollections. (Dave Taylor from *A Suitcase Full of Dreams*)

Storm damage 18 July 1946 – SZ559 in the foreground and RN264 in the background. Ken Delve Collection

One of the most unusual Sunderlands; Dundas Bednal, CO of 230 Squadron, had JM673 painted in matt black: the aircraft become known as 'Black Peter'. Dundas Bednal via Arthur Banks

The Japan transport service was terminated in April 1948; during the period of this service 88 Squadron had carried 2,368 passengers, 202,779lb of mail and 200,270lb of freight. One of the standard tasks for the Sunderlands of the FEFBW (Far East Flying Boat Wing) was acting as ASR cover when single-seat fighters were routing to and from Hong Kong; on 11 January the ASR cover being provided for Vampires routing from Singapore to Hong Kong, via Saigon, was called on to act when one of the fighters ran out of fuel and had to force-land on the beach at Bias Bay – in Chinese territory. Sunderland 'D' located the aircraft and passed the position to HMS *Belfast*. The following day two Sunderlands flew a technical party and equipment to Bias Bay, the intention being to salvage the aircraft and put it on the warship; this was successfully accomplished and the Sunderlands flew the technical party back out.

One of the main roles for the flying boats was, however, that of Search and Rescue and an aircraft was kept at one-hour readiness at Seletar, with others on standby at different locations if required, Hong Kong and Trincomalee (Ceylon) being routine locations. Whilst on SAR duty the aircraft was equipped with Lindholme Gear.

Operation *Firedog*

After a reasonably peaceful and settled 1947, the Far East theatre was about to enter a four-year period which saw the RAF, and the Sunderlands, involved in a number of 'active operations'. During the war the British had armed, trained and encouraged the anti-Japanese communist guerillas in Malaya. The Malayan Races Liberation Army (MRLA) began operations against the British in 1948 – the Malayan Emergency was underway and, as Operation *Firedog*, various RAF units would undertake offensive operations for the next ten years. Sunderlands were involved almost from the start.

The first recorded sorties by 88 Squadron were flown on 6 September, three aircraft being tasked, Sqn Ldr Gall and Flt Lt Letford completed their missions but the third aircraft had to return early with electrical problems.

While involvement in the Malayan Emergency was still building up, 88 Squadron hit the headlines with its involvement in the 'Yangtse Incident'.

The Yangtse Incident

'British warships shelled on Yangtse' blazed the headlines of *The Times* on Thursday 21 April, 1949. On 20 April the frigate HMS *Amethyst* was steaming up river from Shanghai to Nanking when she was fired upon by communist Chinese shore batteries in the vicinity of Chinkiang. *Amethyst* had been on her way to replace HMS *Consort* as the Shanghai guard ship, a part of the British presence to protect Western 'interests' in China during the communist advance towards Shanghai (civil war having broken out in China).

At 1130 on 20 April, Air Headquarters (AHQ) Hong Kong sent a warning order to 88 Squadron at Kai Tak to bring the Squadron to readiness, to have all crews available and all aircraft serviceable. The squadron was eager to go. At 2225 the same day a more specific request was received, for a single aircraft to drop supplies to a party which had left the ship and was making its way overland towards Shanghai. One aircraft was made ready – fully armed and stocked with food, water and medical supplies plus, of course, para dropping gear. A crew was nominated and two doctors were added to the normal complement. Next day dawned with the situation unchanged; the damaged *Amethyst* lay aground on Rose Island, fifteen miles east of Chinkiang, whilst the relief force of HMSS *Consort*, *London* and *Black Swan* sat at Kiangyin unable to make any progress. *Consort* had already had one tangle with the gun batteries and had suffered seventy-two casualties. Later on 21 April, *London* and *Black Swan* attempted to get through to the stranded frigate but the gun batteries were ready and both vessels took damage and casualties before abandoning the attempt.

At 2225 Zulu (GMT), Sunderland ML772/D of 88 Squadron took off from it base at Hong Kong under the command of Flt Lt Ken Letford and, after an uneventful flight, was approaching the *Amethyst*'s position. When about 5 miles away the crew received a priority message from HMS *London* not to approach the *Amethyst* as gunfire was too intense, but to proceed to Shanghai and await orders. An hour later (0405 Zulu) the aircraft landed at Lung Wha, on the Whang Poo river, Shanghai.

The situation on board *Amethyst* was serious, with a number of wounded, no morphia and very little medical care. Medical supplies and a doctor were urgently needed. A request was made by Flag Officer Second in Command Far East Fleet for the Sunderland to land alongside the frigate to transfer a doctor and medical supplies. Intelligence reports indicated that the south bank of the Yangtse was held by Chinese Nationalist forces, with the north bank in the hands of the communists. This meant that the hostile gun batteries were close to any landing site and the Sunderland would almost certainly come under fire. So, at 0805, Letford and his crew again set off for the *Amethyst*, the plan being to land in a nearby creek and transfer the supplies by sampan.

However, when the Sunderland arrived and overflew the stricken vessel there was no hostile reception. Letford therefore elected to land close to the *Amethyst* to save time. He set the Sunderland down on the river and taxied towards the frigate as a sampan came out from the ship to meet the aircraft. At that moment the shore batteries opened fire with a heavy and accurate fire. One doctor, Flt Lt Fearnley, and medical supplies had been transferred to the sampan and a naval officer from *Amethyst* had boarded the Sunderland. Before the naval officer could get back into the sampan it left, and as an immediate take-off was essential in order to save the Sunderland, he had to remain on the aircraft. The anchor couldn't be raised so Letford ordered the cable to be cut. The shellfire was such that he had to take-off downwind and with the tide, and the Sunderland took three miles to get airborne! At any stage disaster could have struck as the river surface had not been cleared and any floating debris could have holed the aircraft's hull.

Yet another conference took place on HMS *London* that night. One thing was clear from the messages sent by *Amethyst*: the Sunderland flight had done wonders for

Sunderland ML772 of 88 Squadron on the Whang Poo river, Shanghai, April 1949. Ken Delve Collection

the ship's morale in that, at last, the crew had visible signs that help was on its way. At Lung Wha the Sunderland detachment was supported by the BOAC flying boat organization and the local agents, Jardine Matheson. The crew was accommodated in the Palace Hotel with Room 305 being used as an Ops Room! This excellent accommodation, and drinking at Marcos, made the whole escapade a little unreal. The only limitation on the social life was a 10pm curfew.

Flt Lt Ken Letford DSO DFC along with part of the crew involved in the *Amethyst* **flights.** Ken Delve Collection

The morning of 22 April brought a request for another flight, this time to transfer eight naval personnel and a naval chaplain. With Letford once more at the controls, the Sunderland was airborne at 0420. *Amethyst* had moved five miles down river but the artillery had moved with her. After three overflights had brought no response from the communist forces, the Sunderland touched down on the river and made towards the ship. As it did so, an accurate barrage of artillery and small arms began and the aircraft was hit. No sampan was available so it was put to the naval personnel that if they were urgently needed perhaps they might care to swim for it! But by then the current had carried the Sunderland past the frigate and so this option was no longer possible. Minus the wireless aerial, which had been shot off, the aircraft took off. Near Chingkiang the crew spotted two sloops and then saw a Mosquito and six Mustangs. Being unsure of anyone's intentions, Letford descended to 200ft (60m) to return to Shanghai at low level. Shortly afterwards the Sunderland was fired on from the ground and was hit at least twice, one bullet passing through a crewman's sleeve. As the aircraft climbed, the crew spotted an aircraft closing in from behind. Letford tried to turn into it but the twin-engined pursuer followed the move. Fortunately, Letford was able to dive into cloud to evade the unknown aircraft. The Sunderland returned to Shanghai, landing at 0730.

The remainder of the day passed quietly but in the evening a second aircraft arrived, NJ176/F, flown by Fg Off Dulieu. Saturday saw another sortie up the Yangtse, this time for a reconnaissance of the river to note the position of any Nationalist ships. The intention was to try to slip the *Amethyst* out with any Nationalist ships that might be trying to make their escape from the river. After take-off the Sunderland flew just below the cloudbase of 1,000ft (300m) en route to the former Nationalist naval base at Chieng-Yin. However, 5 miles south of this position the aircraft was hit in the port main fuel tank, petrol gushing out and filling the aircraft with fumes. Also, the hydraulics of the rear turret had been damaged and the turret was out of action. The Sunderland went into evasive manoeuvres but was unable to return fire as no distinct targets could be seen. The damage required an immediate return to Shanghai, where the aircraft landed at 0355.

Sunderland ML772 was now out of the battle as the damage it had suffered required its return to Hong Kong for repairs. On Sunday 24 April the aircraft left Shanghai on its way home and NJ176 became the primary aircraft. However, Letford and his crew stayed with the replacement aircraft as they were now considered the best crew for the job in the light of their recent experience. On the same day, intelligence reports showed that fighting had passed the *Amethyst* as the communist advance towards Shanghai continued. *Amethyst* was now stranded in hostile territory and on 25 April the Navy stated that it had no further use for the Sunderland as it was evacuating Shanghai. The detachment commander, Gp Capt J. N. Jefferson, Commanding Officer of Kai Tak, discussed the evacuation plans and offered air evacuation if required. However, at 0510 Sunderland NJ176 left Shanghai and was back at Kai Tak at 1040.

It had been a short 'campaign' for the lone Sunderland of 88 Squadron but the morale boost for the trapped crew on *Amethyst* was enormous. Ken Letford and his crew had carried out a number of hazardous missions: for his inspired leadership, Letford was awarded a bar to his DFC. For *Amethyst* the story was not yet over. After the April engagement the plan was for the ship to 'sit it out' until a negotiated settlement could be reached. However, the communist authorities demanded an apology from Britain for 'wrongful invasion' and compensation for the villages shelled during the conflict. Britain said that nothing could be done until the ship was freed. During the stalemate the ship's crew was allowed to purchase limited amounts of food locally, but during the early summer problems with rats and mosquitoes made life unbearable on board. The steady consumption of irreplaceable fuel supplies placed an effective deadline on any escape attempt. On a Saturday evening in late July, *Amethyst* slipped her moorings and made for the open sea. Despite opposition from various shore batteries, the 140-mile (225km) journey was made in just over

7 hours – with no pilot and sailing by night. The Yangtse incident was over.

> News was received at noon that the evacuation of British nationals from Shanghai was definitely to take place. A large briefing was held in the afternoon and it was decided that aircraft on arrival should receive a visual green from the ground or launch before alighting. Although it was known that Communist Forces were closing in on Shanghai and Lung Wha, the speed of the advance was unpredictable. All available aircraft were refuelled and standing by until it was know that only one aircraft was required the following day and possibly three the day after. (88 Squadron ORB, May 1949)

At 0400 on 15 May Sqn Ldr Gall left Kai Tak in 'F' and landed at Shanghai five hours later; twenty-six passengers were picked up and flown back to Kai Tak. Three sorties were flown on the 16th, a total of eighty-three passengers being brought out of Shanghai. The squadron flew numerous trips to Shanghai evacuating British personnel as the communist forces continued their advance.

The Sunderlands were kept active with exercises and various detachments, aircraft operating from a variety of locations, including Augusta, Korangi Creek, Fanarar, Seletar and Glugor, as well as the odd detachment to places farther afield such as New Zealand, the Philippines and Japan. The latter, operating from Iwakuni, was to become a major commitment.

Firedog Hots Up

In addition to the bombing sorties, the Sunderlands from all three squadrons also conducted coastal patrols in co-operation with Royal Navy vessels.

> The patrol of coastal waters by the Far East Flying Boat Wing, islands off the east coast of Malaya had always been a medium through which seaborne illegal entries had been made and when the Emergency broke out it was felt that the dangers of this immigration and the smuggling of arms and equipment to the terrorists by sea had greatly increased. (AP3410, RAF in the Malayan Emergency)

A daily patrol had been mounted by 209 Squadron since June 1948 along the east coast to report any movement of shipping within harbours and within 40 miles of the coast.

The move of 209 Squadron from Ceylon to Seletar on 3 September 1949 was part of the build-up of air assets, giving the AOC a total strike force of ten Sunderlands, seventeen Spitfires, sixteen Tempests and eight Beaufighters.

However, with the departure of 209 Squadron to Iwakuni, Japan, in September 1950 to participate in the Korean War, the Sunderland force was reduced to the five aircraft of 205 Squadron: 'the loss of 209 Squadron was potential rather than actual, however, as the type of action for which they were best suited was rare at this stage of the campaign' (AP3410). The temporary return of 88 Squadron from Hong Kong the following year provided a boost to the Sunderland force. Typical of the major anti-terrorist operations carried out was Operation *Sword*, which ran from July 1953 to March 1954, covering an area of the Bongsu Forest Reserve. A series of jungle areas, including all know CT camps, were chosen as targets and bombed by Lincolns (on detachment from Bomber Command), Sunderlands and Hornets.

> Protracted and haphazard blasting of the area by Sunderlands armed with small anti-personnel bombs was carried out during the 24 hours or 48 hours following a major air strike in order to harass the survivors and create the greatest feeling of anxiety and uncertainty. … Even more effective in the harassing role was the 20lb fragmentation bomb with a mean area of effectiveness of 1,000sq ft, that was introduced into the campaign in 1949. The noise of its detonation was not appreciably less than that of a 500lb bomb and as a recipient got sixteen instead of one, its effect on morale, if not its lethal power, was greater. The 20lb bomb was therefore a weapon of great potential but as the Sunderland alone was capable of carrying sufficient quantities to justify its employment, it was not used in great quantities and none were dropped after these aircraft were withdrawn in 1958. (AP3410)

The only Sunderland 'operational' loss during this period was SZ573, the aircraft being destroyed during bombing-up on 26 March 1950.

Sunderland GR.5 involvement in Operation *Firedog* comprised:

Unit	Dates
88 Squadron	1/10/50–5/4/54
205 Squadron	15/9/49–13/12/54
209 Squadron	1/1/48–31/12/54
205/209 Squadron	31/12/54–1/11/58

1955 brought the amalgamation of 205 and 209 Squadrons, as 205 Squadron, with eight aircraft, this unit remaining operational on *Firedog* sorties until March 1958.

The Sunderlands were called on to act as bombers in the anti-terrorist operations. As Colin Sharpe recalled:

> Our method of bombing was crude in the extreme. We carried 25lb [sic] bombs (two in a box) which seemed to fill every available space in the aircraft – the bomb room, the galley, the wardroom and the aft areas. When we reached the target area and it was time for the drop, the boxes were unpacked, the bombs taken out and the firing pin removed from the small propeller on the nose of the bomb to enable the bomb to arm itself on the way down. The bombs were then thrown out of the galley windows on either side of the aircraft after a call by the pilot. Maybe six or eight bombs could be dropped on each run, after which we would turn and fly back on a reciprocal course to repeat the process. We would remain over the target for perhaps four or five hours, dropping bombs and observing the effects. Obviously many of the bombs exploded when they hit the rain forest canopy but presumably it wasn't very pleasant for the CTs to have shrapnel whistling about them.

The main advantage of the Sunderland in such bombing operations was its on-target endurance, lurking for hours and dropping a series of anti-personnel bombs. Flt Lt Hunter flew 88 Squadron's first night *Firedog* on 17 January, the Sunderland making a timed run over two searchlights and dropping the standard load of 160 × 20lb fragmentation bombs.

Colin Sharpe again:

> Night *Firedogs* were a different matter altogether – and much more demanding. During these operations we would be in radio contact with the ground controller, who would set up two jeeps with crossed searchlight beams to show the start point. We would then use this reference to attack the target area. The usual dropping height was 1,000ft above ground level and so night *Firedogs* in the mountains of central and northern Malaya were not the most pleasant operational tasks to perform! They required very accurate timing and precision flying as the target area might be only 5 miles wide by 10 miles long and was often in a valley with mountains all around.

The squadron's ORB for July 1951 records the intensity of its involvement with

Operation *Firedog*, the following sorties being flown (*see* table right).

There was always time to fly the odd 'public relations' sortie: on 21 September Flt Lt Hunter's crew provided a 'firework display' for Singapore City Day, the Sunderland flying over Kallang Airport dropping anti-submarine flares and firing various colours of Verey cartridge.

Flt Lt G. N. Sims was airborne from Seletar air-testing ML882 of 88 Squadron when he was ordered to proceed towards China Rock where two aircraft had collided:

I immediately set course for the rock and approx. 5 miles from the east coast of Johore I was intercepted by a Mosquito which lowered its wheels and waggled its wings. I followed the Mosquito until we reached a point on the Johore coast from where I could see two areas of burning debris and, slightly to shoreward, two parachutes entangled in trees approx 300yd apart. Circling the area I decided to investigate the beach area and on the first circuit located one survivor clad in flying overalls and waving a white handkerchief. I flew low over the person and fired a green Verey cartridge, and on continuing the circuit encountered 'M' of 205 Squadron, Flt Lt Bridge. As 'Mike' was acting as ASR aircraft he proceeded to effect a landing approx 800yd off shore. The landing was successfully carried out into a swell estimated from the air to be about 2ft.

During this time, I had notified 'Mike' that I would attempt to persuade a junk which was towing timber to cast off its dinghy and take it ashore. This was effected by flying low over the junk in the direction of the survivor and firing a red Verey cartridge. After three such runs the junk master correctly interpreted the manoeuvre and the dinghy made for shore, where the first survivor was rescued.

As soon as the first survivor had been rescued, I conducted a further search along the foreshore to the north and located the second survivor, who had discarded his flying suit to make a ground signal on the beach. On the second circuit of the position, we found that the person had re-entered the jungle but had left the inscription 'OK' in the sand. As by this time my VHF transmitter was unserviceable I passed the following message by Aldis 'Will fire red over second survivor'. On the third run over the position, firing reds each time, I received 'OK' and the junk's dinghy picked up the second survivor.

I was running low on fuel, having one hours' flying time left, so dropping a marker just off shore, I set course for base.

Date	*A/C*	*Captain*	*Hours airborne*	*Bombs and ammunition carried*
1 Jul	'B'	Flt Lt Brand	8hr 50min	160 × 20lb, 5,000 0.303in, 1,000 0.5in
	'C'	Sqn Ldr Helme	4hr 55min	160 × 20lb, 4,500 0.303in, 400 0.5in
2 Jul	'D'	Sgt Weaver	4hr 15min	160 × 20lb, 5,000 0.303in, 1,000 0.5in
	'B'	Flt Lt Douche	5hr 35min	160 × 20lb, 5,000 0.303in, 1,000 0.5in
	'C'	Sqn Ldr Helme	5hr 25min	160 × 20lb, 4,800 0.303in, 1,000 0.5in
3 Jul	'F'	Flt Lt Brand	4hr 35min	160 × 20lb, 2,500 0.303in, 300 0.5in
4 Jul	'B'	Flt Lt Douche	5hr	160 × 20lb, 3,800 0.303in, 1,000 0.5in
	'C'	Sqn Ldr Helme	2hr 55min	160 × 20lb, 4,500 0.303in, 2,000 0.5in
	'F'	Fg Off Boston	3hr 55min	Unknown
5 Jul	'B'	Flt Lt Douche	4hr 45min	160 × 20lb, 3,800 0.303in, 1,000 0.5in
	'F'	Flt Lt Brand	4hr 5min	160 × 20lb, 4,500 0.303in, 600 0.5in
7 Jul	'F'	Flt Lt Douche	7hr 35min	160 × 20lb, 5,000 0.303in, 1,000 0.5in
8 Jul	'F'	Flt Lt Brand	8hr 5min	160 × 20lb, 5,000 0.303in, 550 0.5in
9 Jul	'A'	Sgt Weaver	6hr 30min	160 × 20lb, 5,000 0.303in, 1,000 0.5in
10 Jul	'F'	Flt Lt Hunter	4hr 10min	160 × 20lb, 1,100 0.303in
16 Jul	'A'	Flt Lt Hunter	6hr 5min	232 × 20lb, 4,500 0.303in, 100 0.5in
17 Jul	'A'	Flt Lt Hunter	7hr 5min	160 × 20lb, 5,000 0.303in, 200 0.5in
18 Jul	'A'	Flt Lt Hunter	6hr 20min	232 × 20lb, 5,000 0.303in, 1,800 0.5in
19 Jul	'A'	Flt Lt Hunter	5hr 25min	232 × 20lb, 5,000 0.303in, 1,000 0.5in
20 Jul	'A'	Flt Lt Hunter	6hr	232 × 20lb, 3,600 0.303in, 1,000 0.5in
21 Jul	'A'	Flt Lt Hunter	3hr 5min	232 × 20lb, 3,900 0.303in, 1,000 0.5in
22 Jul	'A'	Flt Lt Houtheusen	5hr 30min	160 × 20lb, 3,500 0.303in, 850 0.5in
24 Jul	'A'	Flt Lt Hunter	6hr 50min	188 × 20lb, 3,300 0.303in, 700 0.5in
25 Jul	'A'	Sgt Weaver	7hr 10min	160 × 20lb, 4,000 0.303in, 1,000 0.5in

Colin Sharpe amassed over 2,000 hours on Sunderlands with the Far East Flying Boat Wing; after a maritime course at St Mawgan (using Lancaster MR.3s) he was posted to Calshot for the Sunderland GR.5 conversion course:

Whilst on the course I had one rather interesting experience. We had returned, one Saturday afternoon, from a long training detail over the Bay of Biscay. A deep depression was centred over Biscay and when we returned for landing a Force 6 gale was blowing which made our landing interesting to say the least, even in the protected waters of the Solent. I and the Navigator were delayed on board the aircraft after we moored up on the trotts (as they were called). When the tender driver returned for us he had difficulty picking us up as the tide had now turned and was against the wind causing a very short, high chop. Anyway he eventually managed to manoeuvre in and collect us and we were very thankful to reach shore. As we were walking away from the pier I noticed that another of the Sunderlands moored further out was listing at an odd angle. We rushed back to the pier and alerted the tender driver, only one being on duty as it was Saturday afternoon. This was an emergency and we had to do something straight away. We collected all the lead weights kept specifically for the purpose of righting a listing aircraft and set out again on the choppy water towards the aircraft. What had happened was that the port float had detached from the aircraft, perhaps weakened as the result of a heavy landing earlier in the day, and the port wing was sinking into the water. With some difficulty I managed to get in through a window. By this time the chop was so bad that the tender driver moved away for fear of knocking a hole in the aircraft – so I was on board alone. The drill was to carry all the lead weights to the wing with the good float, which would then lift the listing float out of the water. This I did but it became increasingly difficult, and as I did not have enough lead weights the slope of the wing increased, when I had put the last of the lead weights in place I realized that the slope on the wing was such that if I let go it would mean a toboggan ride for me down the wing, over the fuselage and into the sea. I had no choice but to remain where I was, The angle of the wing continued to increase as the other wing took on more water. An American cruise liner passed by – but slow enough not to cause a major wake (although the passengers lined up to take photographs). My Navigator had phoned the duty officer, who collected twenty-two airmen who were playing football at the time, then as the wind had abated they were all able to get on board and throw me a rope which I secured; they then clambered up the wing to join me and the combined weight was enough for the aircraft to right itself. It was towed back and repaired.

Sunderland participation in the Korean War was not on a large scale but it did provide an additional aspect of air power. Here 'C' begins its take-off run, adjacent to its operations control ship, in the Iwakuni Roads. Ken Delve Collection

Korean War

On 25 June 1950, North Korean forces crossed the 38th Parallel in an attempt to seize control of the entire country. Under American leadership this was countered by a United Nations force; air power played a key role in this conflict and the role of the Sunderland was small but not insignificant.

Sunderland RN277 of 88 Squadron was the first of the Squadron's aircraft to deploy to Iwakuni, Japan (June 1950) as part of the Allied air build-up. Throughout the summer the Far East Flying Boat Wing maintained a detachment of between three and six aircraft in Japan, the primary task being night operations in the Yellow Sea and reconnaissance sorties along the coasts of Korea. Anti-submarine patrols were flown to cover refuelling operations by US Naval vessels.

On 5 October 88 Squadron deployed one aircraft to Inchon, the UN forces having conducted a bold amphibious landing at this location.

Operations supporting Allied naval forces off Korea occupied most flying hours in January, the squadron recording the following sorties (*see* table right).

Date	*Aircraft*	*Captain*	*Time*	*Details*
1 Jan	ML745	Fg Off Douche	6hr 45min	ASP off east coast of Korea
4 Jan	PP155	Fg Off Brand	9hr 40min	Wx recce east coast of Korea
5 Jan	ML745	Fg Off Douche	11hr 50min	ASP off east coast of Korea
8 Jan	PP155	Fg Off Brand	8hr	ASP Tsushima Straits
	ML745	Fg Off Douche	8hr 25min	ASP support Refuelling Gp
10 Jan	PP155	Fg Off Brand	8hr 45min	Wx recce east coast of Korea
11 Jan	ML745	Fg Off Douche	10hr 10min	ASP support Refuelling Gp
12 Jan	RN282	Flt Lt Houtheusen	7hr 55min	ASP support Refuelling Gp
14 Jan	ML745	Fg Off Douche	11hr 20min	ASP Tsushima Straits
	PP155	Fg Off Brand	8hr 45min	ASP support Refuelling Gp
15 Jan	RN282	Flt Lt Houtheusen	11hr 15min	(*see* text below left)
16 Jan	PP155	Fg Off Brand	10hr 10min	ASP support Refuelling Gp
17 Jan	RN282	Flt Lt Houtheusen	12hr 45min	ASP Tsushima Straits
18 Jan	PP155	Fg Off Brand	10hr 15min	ASP support Refuelling Gp
19 Jan	ML745	Fg Off Douche	9hr 10min	Wx recce east coast of Korea
	RN282	Flt Lt Houtheusen	9hr 55min	ASP support Refuelling Gp
22 Jan	ML745	Fg Off Douche	9hr 25min	ASP support Refuelling Gp
24 Jan	ML745	Flt Lt Hunter	8hr 55min	ASP support Refuelling Gp
25 Jan	RN277	Sgt Weaver	10hr	ASP support Refuelling Gp
26 Jan	RN282	Flt Lt Houtheusen	9hr 40min	ASP Tsushima Straits
28 Jan	ML745	Flt Lt Hunter	8hr 10min	Wx recce
30 Jan	RN282	Flt Lt Houtheusen	9hr 10min	ASP support refuelling Gp

On 15 January, having left Iwakuni at 1030 for a standard anti-submarine patrol in support of the Refuelling Group, Sunderland RN282/C (Flt Lt Houtheusen) was notified of a downed US Navy Corsair pilot who was in his dinghy close to the enemy coast. Houtheusen was able to land near the dinghy and pick up the exhausted pilot, who was subsequently treated for exposure during the flight back to Iwakuni.

The majority of sorties were, however, ASPs in the Tsushima Straits. The following day Fg Off Brand was on a similar patrol when the crew spotted a sampan 30 miles (50km) off the coast: closer examination showed that the sampan's crew of some thirty Koreans was displaying an SOS signal and so the Sunderland notified the USS *Norris* to investigate.

On 25 March 1953 Sunderland PP148 crashed when landing in heavy seas at Iwakuni, five of the crew being killed. The pilot, Flt Lt Wilkinson, gave the following account of the accident:

The aircraft arrived over Iwakuni at 1040Z at 5,000ft to clear heavy turbulence. Seaplane control reported 35kt wind, gusting to 50kt – sea state moderate chop. A low run was made at 50ft using landing light to assess the sea state. The control ship reported that the wind was due to slacken off at about 1200Z. I assessed the sea state and air turbulence as being too rough for an immediate alighting, and decided to orbit. At 1200Z, an emergency flare path was laid into wind and the control ship reported the wind was steady at 24kt, gusting to 35kt. Owing to fuel state and consequent lightness of the aircraft, I decided to attempt to land. A normal night-type approach was made, with crew in crash positions. Twice the aircraft was thrown off the water, I overshot and went round for another attempt, using normal day-time approach. I felt

Three Sunderlands at moorings; the nearest aircraft, SZ577, served with most squadrons of the Far East Flying Boat Wing and was finally disposed of in March 1957. Ken Delve Collection

'All aboard' – a lighter brings the crew to an 88 Squadron aircraft. Note the mooring attachments at the front of the turret. Ken Delve Collection

the aircraft touch the wave crests and began normal throttling back. The next thing I knew was that I was under water. When I came to the surface the aircraft was floating on its back, with floats intact and the hull sheared at the galley with the nose section missing.

The year ended with the loss on 27 December of 88 Squadron's RN302/C: whilst on an anti-submarine patrol in the Tsushima Straits, the aircraft suffered an engine fire plus loss of power on another engine.

February 1954 brought a ceasefire in the Korean War – and the announcement of another batch of gallantry awards, including DFCs for Sqn Ldr H. Francis, Flt Lt M. Bennett, and a DFM for Sgt J. Kitching (all from 88 Squadron).

Anti-Piracy Patrols

At the end of the month a joint RAF/RN Anti-Piracy Patrol Order (APPO) was issued:

Acts of piracy are being committed on the High Seas in the Malacca Straits between the Singapore Strait and the One Fathom Bank Light. The pirated craft are fishing vessels from Port Swettenham, Malacca and Kukup. Own forces comprise one destroyer and one frigate plus two 72ft SDMLs along with Sunderland aircraft of the Far East Flying Boat Wing.

The aim is to discourage piracy by the presence of air patrols, to obtain photographs and to vector HM vessels to the area. (Op Order APPO dated 31 October)

The first such patrol had been flown even before the issuing of the APPO, a sortie being flown by EJ155 of 88 Squadron on 24 October. On average each squadron flew two such patrols a month. Detachments operated from Labuan on anti-piracy patrols:

For these operations we would be based in Labuan for the months of the NE Monsoon; at this time the pirates, who came from the southern islands of the Philippines, would take what they probably thought as their summer holidays! They were using diesel-powered boats that were faster than the police boats; however, an aircraft in the area changed the situation. In one day we could cover the complete area, looking up rivers and landing, if necessary, to check out suspicious boats. On the sorties we carried a police inspector who used binoculars to assess if the boat was suspicious. All legitimate fishermen were issued with papers which they would often wave as we made a low pass by them. If we landed then we could pick up the captain and take him back to Labuan for further questioning. Occasionally they would resist arrest – and they were always armed – whereupon we would sink them with cannon and machine-gun fire; the survivors would be picked up by a police boat if one was nearby or we would land and pick them up. (Colin Sharpe)

The end for the Far East Flying Boat Wing was not far off and with the end of the Korean War there were few tasks for the squadrons, although they maintained a comprehensive training programme as it was by no means certain that the cease-fire would hold. One of the major tasks for April was that of providing escort for the Royal Convoy and to this end 88 Squadron deployed five Sunderlands to China Bay, Ceylon in early April. On 8 April the five aircraft flew in salute past the SS *Gothic*, bearing the Royal party, SZ578 remaining with the ship as escort whilst the others flew back to Kai Tak.

1 October 1954 saw 88 Squadron disband at Seletar, aircraft being handed to the remaining two squadrons of the Far East Flying Boat Wing. The following January, these two units were amalgamated (*see* above) as 205 Squadron but this unit, too, was only given a temporary stay of execution.

With the help of a truck and the usual 'tug of war' group of airmen, VB880 is brought up the slip; this aircraft was transferred to the RNZAF in July 1953. Ken Delve Collection

Colin Sharpe recalls:

Early in 1955 my crew was on SAR duty and at around midnight we were called out to carry out a night search of an area around the Great Natuna islands in the South China Sea. This night search involved flying a search pattern based on the visibility in the area at the time and at the end of each search by firing a Verey cartridge and then awaiting a flare from a dinghy if one existed. At dawn we went back to the main island to update our plot and found the remains of an aircraft in Small Bay – but only a wheel, a patch of oil and assorted debris. The aircraft was Air India Constellation 1049 *Kashmir Princess* which I had seen take off two days earlier from Kallang. The

***(Right)* Part of the Sunderland escort provided by 8 Squadron for HMS** Gothic **during the Royal Visit of April 1945.** Ken Delve Collection

aircraft was carrying a Communist Chinese delegation to the non-aligned countries conference in Indonesia and it was widely rumoured that it had been blown up by a bomb.

Operating primarily in the transport role, but at the same time maintaining its maritime skills, 205 Squadron carried the banner as the RAF's last Sunderland squadron. Shackleton MR.1As began to arrive in May 1958 but the Sunderland held on for another year, the last 'operational' sortie – an exercise – being flown by DP1987 on 14 May. With the demise of 205 Squadron's Sunderlands, the RAF Sunderland story came to an end.

209 Squadron's VB688 at Seletar, 1953.
Peter Green Collection

***(Below)* Disbandment parade for 88 Squadron at Seletar on 2 October 1954. JM667 provides the backdrop – and was struck off charge a few days later.** Peter Green Collection

CHAPTER EIGHT

The Civilians

This chapter provides a brief overview of Shorts civilian flying boats from the C-Class to various Sunderland derivatives.

On 1 July 1936 the first of forty-two Empire flying boats emerged from No. 3 Erecting Shop at Rochester, G-ADHL 'Canopus' duly making its first flight on 4 July. 'Canopus' commenced revenue service in early November as the Imperial Airways C-Class flagship, operating in the Mediterranean region. In July 1936 Imperial Airways had created four operational divisions for the Empire Air Mail Service (EAMS), although it was June 1938 before all of these were fully operational:

No.1 Division – Europe.
No.2 Division – Mediterranean and Africa (east to Karachi, south to Kisumu).
No.3 Division – Karachi-Singapore-Hong Kong.
No.4 Division – South of Kisumu.

The remaining C-Class boats were delivered at two aircraft per month in 1937 and services were soon operating to the Far East and Africa, and from May 1937 a Bermuda–New York service was in operation. Several aircraft underwent modification for mail services, 'Caledonia', for example, having its fuel capacity increased (with a 45,000lb AUW) for direct operations across the North Atlantic. However, the details of the C- Class/Empire boat operations are beyond the scope of this book. It is worthy of note that a number of Empire boats survived the war, the final services being flown in the latter part of 1947, after which these classic aircraft met the fate of being broken up on Southampton Water.

With the outbreak of war BOAC's passengers changed from being wealthy tourists and business people to senior RAF officers, War Office personnel, and aircrew. The war also saw the end of EAMS.

Under wartime requirements the interior fitting of the aircraft was modified, the number of seats being increased, giving higher take-off weights and so a need for more power from the engines.

The entry of Italy into the war in mid-1940 meant a change for BOAC in terms of its routes and the basing of its flying boats. A station was established at Durban, South Africa for the so-called 'Horseshoe' route connecting South Africa, East Africa, Cairo, Iraq, India, Burma, Malaya and Singapore.

The BOAC bosses decided that the Sunderland held more promise than the old C-Class boats and BOAC acquired twelve Sunderland IIIs from the RAF in the latter part of 1942, armament having

C-Class boat G-AEUF 'Corinthian'; the C-Class Empire boats entered service in 1936. FlyPast Collection

been deleted but camouflage retained. The aircraft were tasked with carriage of priority freight (and passengers), primarily on the route UK to West Africa, and by early 1943 had begun to acquire civil serial numbers (G-AGER to G-AGEW). Additional Sunderlands were acquired at various times.

Jack Harrington was given the task of finding suitable pilots for the new aircraft; Howard Fry was one of those he approached:

> Jack Harrington explained the situation to us: 'Look chaps, the RAF have given us this fleet of Sunderlands and they want a daily service, through Foynes, Lisbon, Bathurst and Freetown down to Lagos. So I've called in all the boat pilots I can find. We're laying on a very brief refresher course on boats; giving you a very quick run around the Sunderland so you'll know the differences from the old C-Class. Then it's going to be one run down to Lagos and back. And then you'll be on your own.'
>
> Raymond Winn gave me about six hours dual on the Sunderland and then I did my trip down the route with Reg Hallam, an ex-RAF man. I found the Sunderland a great improvement on the older boats. The longer nose made it easier to judge the landing and reduced the amount of porpoising on a swell. Then there were big improvements in the engines.

The BOAC Sunderlands were equipped with thirty-nine seats and none of the 'frills' found on a civilian aircraft; the gun turrets were removed and the RAF roundels painted out. However, when G-AGER was flown to Durban for evaluation, the Station Manager was not impressed and turned down the type.

	Sunderland	**C-Class**	**G-Class**
Span:	112ft 9½in	114ft	134ft 4in
Length:	85ft 4in	88ft	103ft 2in
Height:	32ft 10½in	31ft 9¾in	37ft 7in
Engines:	Pegasus XXII	Pegasus XC	Hercules IV
Max speed:	210mph	200mph	214mph
Range:	1,780 miles	810 miles	3,200 miles

Kasfareet 1944; BOAC Sunderland III G-AGEW (JM665). Andy Thomas Collection

G-AGEU (JM663) at Penang in 1946. Peter Green Collection

Short C-Class Derivatives

The Modified C-Class flying boats were powered by Perseus XIIC engines and with various airframe strengthening had a loaded weight of 48,000lb (22,000kg), which could be increased to 53,000lb (24,000kg) with air refuelling.

G-Class

A small number of G-Class boats, powered by four Bristol Hercules IVC engines, was built, the first of which, 'Golden Hind' (G-AFCI), made its maiden flight in June 1939. The intention was for these aircraft to operate across the Atlantic, but with the outbreak of war they were commandeered by the Air Ministry and, in August 1940, allocated to 119 Squadron. Two of the four aircraft survived their military service and returned to BOAC in 1942, operating a service from Poole to West Africa. 'Golden Horn' crashed in January 1943 and the sole survivor, 'Golden Hind', was subsequently converted to Hythe standard.

G-Class boat G-AFCI with BOAC at Koggala, probably late 1945. Peter Green Collection

(Below) **Very much a Sunderland with faired-over gun turrets, but with an improved interior – Hythe G-AGJM.** Ken Delve Collection

Hythe (S.25)

With the end of the war in Europe the civilianized Sunderlands were put through a conversion (and de-camouflaging) programme at Hythe and Belfast to suit them for service with BOAC on the old Empire routes. With new engines, Pegasus 38s or 48s, and new interiors they were designated as Short Hythes, twenty-four aircraft being converted, the first of which began service in January 1946.

Eric 'Timber' Woods joined BOAC in 1946, having flown bombers with the RAF during the war, and was sent to fly Hythe flying boats:

> The Hythe had a pretty good record for serviceability but there were occasions when, for one reason or another, an engine had to be shut down, followed usually by a return to the departure point. Although these aircraft could fly comfortably on three engines, the loss of an engine still led to quite a reduction in speed, which did not help when one was anxious to get back on to safe water again as soon as possible. One one occasion in summer 1947 whilst on our way east from Basra we had the misfortune to have both port engines malfunction, with a need to shut them down. This was disconcerting to say the least, and we were a much relieved crew when G-AGHW 'Hamilton' was thumped somewhat unceremoniously down on the Shatt al Arab river. (From *Flying Boats to Flying Jets*, Eric 'Timber' Woods, Airlife 1997)

The Hythes provided a vital service pending the arrival of newer types to fill the growing requirements of civil aviation; they were phased out in the late 1940s, being replaced by land-based Lockheed Constellations with BOAC.

Sandringham (S.25)

One of the civilianized Sunderlands, G-AGKX 'Himalaya', was sent back to the Rochester works in 1945 for a more comprehensive re-work to create a more suitable transport version. This sole Sandringham I was launched in November 1945 with a substantially modified front end – a more aerodynamic shape – and a purpose-designed interior for twenty-two passengers on two decks. Although no new aircraft were built to this design, a number of

Sunderland Vs were duly converted in the late 1940s, all powered by Pratt & Whitney R-1830s but with various designations (Sandringham 2, 3, 5 or 7) depending on the internal fit. BOAC was the main operator, but the type also served with a number of overseas operators such as Dodero (Argentina), Tasman Empire Airways, Norwegian Air Lines and Aquila Airways.

Solent (S.45)

Another conversion of a military flying boat, the Solent came out of the Seaford (Sunderland IV), the prototype, G-AHIL 'Salisbury', being launched on 11 November 1946. BOAC ordered twelve aircraft powered by four Bristol Hercules 637s and capable of carrying thirty passengers on two decks. As the Solent 2, the type entered service in May 1948 on the London to South Africa route. It was a short-lived period with BOAC as the airline ceased all flying boat operations in 1950. A number of Solents went to other operators, such as Trans-Oceanic Airways in Australia. Six Seafords were altered during production to become Solent 3s with the same engines as the Solent 2 but with a greater endurance, these too going initially to BOAC. The final version, with another increase in endurance but also with an engine change to the Hercules 733, was the Solent 4; this impressive long-range version could carry forty-four passengers. Aquila Airways operated Solents from the UK until late 1958.

***(Below)* Sandringham prototype ML788 on the Medway 29 November 1945; this was the first truly civil aircraft built by Shorts since the outbreak of the war.** Peter Green Collection

***(Bottom)* Short Solent G-AHIL Salisbury was the prototype of this civil conversion from the Seaford.** FlyPast Collection

Sunderland Production

SD – Shot Down
Conv – Converted
DBR – Damaged Beyond Repair
FTR – Failed to Return
Xxd – Crashed
SOC – Struck Off Charge
Ditch – Ditched
Dest – Destroyed
Med – Mediterranean
Wfu – Withdrawn from use

The best single source of information concerning serials and details of RAF aircraft is the Air-Britain series of publications. These books contain details extracted from the RAF Aircraft Record cards.

Aircraft	Units	Fate
Sunderland I – Short Bros (11)		
L2158	MAEE, 204	FTR Sierra Leone 17 Aug 42
L2159	MAEE, 230	DBR Greenock 7 May 41
L2160	230, 4 OTU	To 3372M
L2161	230	Dest Ju 87 Scaramanga 23 Apr 41
L2162	210	Xxd Milford Bay 20 Sep 38
L2163	210, 240, 10 RAAF, 228	Sank Stranraer 15 Jan 42
L2164	230	Sank St Paul's Bay 10 Mar 41
L2165	210	Xxd Milford Haven 18 Sep 39
L2166	230	SD 1 Aug 41
L2167	210	FTR Oslofjord 1 Aug 41
L2168	210, 201, 228, 4 OTU	Xxd nr Nigg 21 Nov 43
Sunderland I – Short Bros (10)		
L5798	210, 201, 204	DBR Gibraltar 20 Sep 43
L5799	210, 204	Xxd nr Bergen 7 Apr 40
L5800	210, 201, 204, 4 OTU	SOC 2 Jul 44
L5801	230	Xxd Johore Straits 5 Jun 39
L5802	210, 204, 201, 95, 461, 4 OTU	Xxd Alness 16 Jan 43
L5803	230, 228, 95, 204, 95	SOC 22 Aug 42
L5804	230	Sank Scaramanga 25 Feb 41
L5805	228, 201, 95	Xxd Atlantic 8 Oct 42
L5806	228, 210, 228, 230	FTR Med 25 Jul 42
L5807	228	Dest Bf 109 Kalafrana 27 Apr 41
Sunderland I – Short Bros (18)		
N9020	228	FTR Ionian Sea 1 Nov 40
N9021	MAEE, 201, 204	Xxd Invergordon 15 Dec 40
N9022	210, 204, 210	Xxd Oban 27 Dec 40
N9023	228, 204	Xxd Iceland 24 Apr 41
N9024	210, 204, 210, 204, 95, 204, 4 OTU	SOC 16 Aug 44
N9025	228, 210, 228	SD by CR42 6 Aug 40
N9026	210	FTR 29 Jun 40
N9027	228, 210, 95,202	Sank Gibraltar 16 Feb 41
N9028	204	FTR 21 Jul 40
N9029	230	FTR 1 Jan 43
N9030	204	Xxd Mount Batten 15 Oct 39
N9044	204, 4 OTU	SOC 24 Jul 44
N9045	204, 4 OTU	Ditch Scilly Islands 13 Oct 39
N9046	204	DBR Sullom Voe 11 Dec 40
N9047	204	Sank Reykjavik 10 Jun 41
N9048	210, 10 RAAF	Dest Mount Batten 28 Nov 40
N9049	10 RAAF	Sank Malta 10 May 41
N9050	210, 10 RAAF, 95, 202, 95, 4 OTU	SOC 29 Jul 44

Attempting to salvage N9049 of 10 Squadron RAAF, Malta, June 1941. Ken Delve Collection

Aircraft	Units	Fate
Sunderland I – Short Bros (12)		
P9600	210, 10 RAAF, 228, 4 OTU	SOC 11 Dec 46
P9601	10 RAAF	Bombed Mount Batten 28 Nov 40
P9602	210, 10 RAAF	DBR Oban 2 Sep 40
P9603	10 RAAF	Xxd Pembroke Dock 26 Jun 41
P9604	10 RAAF, MAEE, 4 OTU	Xxd Wig Bay 11 Jun 42
P9605	10 RAAF, 4 OTU	SOC 10 May 44
P9606	10 RAAF, 201, 4 OTU	SOC 10 May 44
P9620	204	Ditch 29 Oct 40
P9621	228, 201	DBR Scalasaig Bay 9 Oct 40
P9622	228, 201	Xxd Dunnet Head 29 Oct 40
P9623	210, 95	Interned Portugal 14 Feb 41
P9624	210	Xxd Oban 15 Mar 41
Sunderland I – Short Bros (20)		
T9040	204, 95, 202, 95, 4 OTU	DBR Alness 2 Jul 44
T9041	210, 95, 201, 204, 201, 204	Ditch 28 Jun 42
T9042	MAEE, 4 OTU	SOC 20 Feb 45
T9043	210	FTR 2 Sep 40
T9044	210	Sank Pembroke Dock 21 Nov 40
T9045	204	Sank 29 Oct 40
T9046	201, 228, 95	Sank Kalafrana 23 Feb 42
T9047	10 RAAF	DBR 9 Jul 41
T9048	204, 228	Xxd Kalamata 25 Apr 41
T9049	204, 201, 4 OTU	Scuttled Wig Bay 11 Dec 46
T9050	230	Xxd Aboukir 30 Sep 41
T9070	204	Sank Half Die 16 Aug 42
T9071	10 RAAF, 230	Xxd Ras Amr 22 Dec 41
T9072	204, 10 RAAF	Xxd nr Holyhead 5 Dec 41
T9073	210, 95	Sank Wig Bay 22 Jun 44
T9074	210, 204, 95, 204	SOC 25 Oct 43
T9075	210, 10 RAAF	Ditched 29 Apr 41
T9076	210, 201, 4 OTU	SOC 11 Jul 44
T9077	201, 228, 4 OTU	SOC 29 Jul 44
T9078	95, 4 OTU	SOC 24 Jul 44
Sunderland II – Shorts (23)		
W3976	MAEE	Xxd Gareloch 28 Nov 41
W3977	201	Xxd off Donegal 5 Feb 42
W3978	204, 201	Xxd Sullom Voe 11 Aug 41
W3979	10 RAAF	Sank St Govan's Head 1 Mar 42
W3980	201, 4 OTU	4908M Nov 44
W3981	204, 201, 240, 4 OTU	SOC 1 Jul 44

Aircraft	Units	Fate
W3982	201	FTR 21 Aug 41
W3983	10 RAAF, 202, 10 RAAF	4603M Nov 44
W3984	10 RAAF	SOC 20 Jun 44
W3985	10 RAAF	FTR 18 Aug 43
W3986	10 RAAF, 228, 10 RAAF	Xxd Mount Batten 20 May 43
W3987	201, 230	Xxd Aboukir 7 Sep 42
W3988	201	Xxd Carrowmore Pt 3 Dec 41
W3989	202, 228, 202, 228, 4 OTU, 302 FTU	SOC 23 Feb 45
W3990	228, 202, 228, 4 OTU	Sank Alness 15 Feb 43
W3991	228	5016M Jan 45
W3992	228, 4 OTU	Sank Alness 13 Feb 43
W3993	10 RAAF	DBR, SOC 3 Jul 44
W3994	10 RAAF, 202, 10 RAAF	FTR Spanish Coast 30 Jul 42
W3995	228	Sank Lough Erne 11 Jan 43
W3996	228	Bombed Kalafrana 2 Feb 42
W3997	201, 10 RAAF, 201, 4 OTU	Xxd Lossiemouth 12 Oct 43
W3998	201	Xxd Mount Batten 20 Dec 41
Sunderland III – Shorts (27)		
W3999	10 RAAF	SD Bay of Biscay 21 Jun 42
W4000	201	Ditched 1 Aug 42
W4001	201	Sank Castle Archdale 4 Oct 42
W4002	201, 302 FTU	SOC 22 Feb 45
W4003	10 RAAF, 201, 302 FTU	SOC 16 Nov 45
W4004	10 RAAF, 202, 10 RAAF	FTR Bay of Biscay 17 May 43
W4017	228, 302 FTU	SOC 19 Jan 45
W4018	201	Sold Jul 45
W4019	10 RAAF	FTR Bay of Biscay 9 Aug 42
W4020	10 RAAF	Xxd Bay of Biscay 2 Aug 43
W4021	230	Xxd Castellorosso 7 Nov 43
W4022	230	SOC 3 Jan 45
W4023	230	SOC 29 Aug 46
W4024	202, 119, 10 RAAF	Sold Mar 47
W4025	201	SD by convoy 31 Jul 42
W4026	228	Xxd Dunbeath 25 Aug 42
W4027	4 OTU	DBR Alness 5 Mar 44
W4028	202, 119, 4 OTU	SOC 12 Jul 45
W4029	202	Xxd Gibraltar 21 Aug 42
W4030	202, 119, 10 RAAF	DBR Mount Batten 16 Sep 44
W4031	4 OTU	SOC 12 Jul 45
W4032	228	Ditched Tiree 4 Sep 42
W4033	4 OTU	SOC 28 Feb 45
W4034	4 OTU	Sold Feb 47
W4035	4 OTU	Xxd Invergordon 11 Aug 42
W4036	201	Xxd Castle Archdale 18 Nov 43
W4037	202, 4 OTU	For conv to Sandringham
Sunderland II – Blackburn (5)		
W6000	423	Sank Wig Bay 13 Dec 42
W6001	423, 119, 4 OTU	Sank Alness 17 Apr 43
W6002	202, 119, 4 OTU	DBR 26 Jun 44
W6003	202	Xxd Gibraltar 12 Aug 42
W6004	228	Sank Stranraer 29 Dec 42
Sunderland III – Blackburn (20)		
W6005	201, 4 OTU	SOC 27 Mar 45
W6006	423, 4 OTU, 423, 4 OTU	Xxd Invergordon 13 Aug 44
W6007	423, 131 OTU	Sold Feb 47
W6008	423	Ditched 12 Mar 44
W6009	423, 4 OTU	Xxd Alness 14 Jan 45

Aircraft	Units	Fate
W6010	201, 4 OTU	Xxd Alness 27 Jul 44
W6011	423, 131 OTU	Sold Feb 47
W6012	204, 4 OTU	SOC 12 Jul 45
W6013	423	Xxd Ballycastle 5 Dec 43
W6014	201, 4 OTU	SOC 12 Jul 45
W6015	204, 95, 4 OTU	Scrap Sep 47
W6016	95	FTR Bay of Biscay 24 Dec 42
W6026	422, 4 OTU	Xxd Alness 22 Mar 45
W6027	422, 330, 4 OTU	SOC 28 Apr 45
W6028	422	Xxd St Angelo 19 Feb 44
W6029	422	Xxd Oban 19 Dec 42
W6030	422, 330	SOC 10 Jul 44
W6031	422	FTR 20 Nov 43
W6032	422, 131 OTU	Scrap Feb 47
W6033	422	Xxd Hvalfjord 28 Sep 43
Sunderland II – Shorts (15)		
W6050	MAEE, 461	4446M Jan 44
W6051	MAEE, 201, 4 OTU	SOC 10 Jul 44
W6052	423, 330	FTR 5 Jun 43
W6053		SOC 30 Apr 45
W6054	10 RAAF	Xxd Mount Batten 12 Nov 42
W6055	201	Scuttled 11 Dec 46
W6056	246, 4 OTU, 131 OTU	4782M
W6057	246	Scuttled 16 Mar 45
W6058	246	4789M
W6059	201, 330	SOC 5 Apr 44
W6060	246, 4 OTU	Xxd Alness 27 Nov 43
W6061	423, 330	SOC 16 Feb 44
W6062	95	SOC 11 Nov 43
W6063	204, 95	Xxd 29 Mar 43
W6064	423, 330, 4 OTU	SOC 13 Nov 44
Sunderland II – Shorts (10)		
W6065	95	Ditched Oporto 5 Mar 43
W6066	246, 422, 131 OTU	Scrap Feb 47
W6067	330	SOC 5 Feb 5
W6068	330, 423, 131 OTU	Scrap Feb 47
W6075	330	Xxd Lough Neagh 12 May 43
W6076	95	SOC 21 Jun 45
W6077	461	SOC 20 Dec 45
W6078	230	SOC 16 Aug 45
W6079	204	Xxd Port Etienne 2 Oct 43
W6080	308 FTU, Flot	SOC 3 Sep 43
Sunderland III – Blackburn (40)		
DD828	423, 201, 4 OTU	SOC 25 Mar 45
DD829	201, 131 OTU	Scrap Mar 47
DD830	228	Xxd Ailsa Craig 3 Feb 43
DD831	422	DBR Bowmore 24 Jan 44
DD832	MAEE, 4 OTU	Scrap Mar 47
DD833	204	DBR Wig Bay 10 Sep 43
DD834	204, 228	Sold Jun 45 (G-AGPT)
DD835	330, 201, 228, 131 OTU, 228	SOC 12 Jul 45
DD836	228	Sank St Marys 24 Jan 44
DD837	228	FTR Biscay 15 May 43
DD838	228, 423, 422, 4 OTU	SOC Jun 45
DD839	4 OTU	Xxd Alness 27 Nov 43
DD840	4 OTU	SOC 26 Mar 47
DD841	4 OTU	Sold Jun 45 (G-AGPY)

Aircraft	Units	Fate
DD842	4 OTU	SOC 12 Jul 45
DD843	330, 423, 131 OTU, 4 OTU	SOC 12 Jul 45
DD844	330, 4 OTU	SOC 28 Apr 45
DD845	246, 422, 141 OTU?	SOC Feb 47
DD846	246, 422	Xxd Clare Island 25 May 43
DD847	228, 131 OTU	Scrap Feb 47
DD848	201	Xxd Brandon Head 22 Aug 43
DD849	423, 131 OTU	SOC Feb 47
DD850	422, 4 OTU	SOC Feb 47
DD851	330, 4 OTU	Xxd Invergordon 26 Nov 44
DD852	10 RAAF	DBR Plymouth Sound 2 Sep 44
DD853	423, 131 OTU, 461	SOC 21 Jun 45
DD854	422, 131 OTU	SOC 23 May 45
DD855	201, 422, 4 OTU	DBR Alness 20 Apr 45
DD856	330, 4 OTU	SOC May 45?
DD857	201	Xxd Castle Archdale 30 Jun 43
DD858	201, 423	Xxd Belfast Lough 23 Oct 43
DD859	423	SD 4 Aug 43
DD860	423	Sold Mar 45 (G-AHEP)
DD861	422	Ditch 3 Sep 43
DD862	423	DBR 31 May 45
DD863	423	Ditch 13 Nov 43
DD864	228	FTR Biscay 14 Dec 43
DD865	10 RAAF, 302 FTU	SOC 16 Aug 45
DD866	461, 302 FTU	SOC 16 Aug 45
DD867	10 RAAF, 423, 131 OTU	SOC 5 Mar 45
Sunderland III – Short Bros (25)		
DP176	119	Ditch Biscay 15 Apr 43
DP177	10 RAAF	FTR Biscay 11 Aug 43
DP178	422, 330, 4 OTU	FTR 14 Mar 45
DP179	119, 10 RAAF	Ditch Biscay 3 Oct 43
DP180	230	SOC 31 Jan 46
DP181	330, 423	Xxd Castle Archdale 11 Nov 43
DP182	204, 343	Ditch S.Atlantic 2 Feb 44
DP183	330	FTR 20 Mar 44
DP184	330, 4 OTU	SOC 12 Jul 45
DP185	201, 4 OTU	SOC 25 Mar 45
DP186	95	Scrap Feb 47
DP187	308 FTU, 343	To Aeronavale 21 Jun 45
DP188	204	Scrap Feb 47
DP189	230	SOC 16 Aug 45
DP190	270	SOC 21 Jun 45
DP191	423, 131 OTU	(Conv Mk V) To RNZAF
DP192	10 RAAF	To RAAF as A26-6 15 Feb 44
DP193	201, 423, 131 OTU	SOC 26 Mar 47
DP194	302 FTU, 95	SOC 21 Jun 45
DP195		Sold 22 Jan 48 (CX-AKR)
DP196	461, 201	SOC 26 Mar 47
DP197	4 OTU	Xxd Croinach 15 Aug 44
DP198	423, 209, 205, 201, 205	(Conv Mk V) SOC 1 Jun 59
DP199	461, 88	(Conv Mk V) SOC 30 Jun 51
DP200	461, 4 OTU, 230	(Conv Mk V) Scrap Oct 57
Sunderland III – Short Bros (25)		
DV956	95, 302 FTU, 95	Ditch 30 Apr 45
DV957	95	Xxd Wellington 3 Sep 42
DV958	202, 119, 228, 10 RAAF	SOC 26 Apr 46
DV959	204	SOC 21 Jun 45
DV960	461, 131 OTU	SOC 31 May 45

Aircraft	Units	Fate
DV961	461, 4 OTU	To 4666M Mar 44
DV962	202, 119, 461	DBR Pembroke Dock 7 Jun 43
DV963	95	SOC 13 Jul 45
DV964	95	Sold 21 Jun 45 (G-AGPZ)
DV965	204, Flot 7E, 343	DBR Wig Bay 21 Sep 45
DV966	204	DBR Wig Bay 21 Sep 45
DV967	MAEE, 228	FTR Biscay 13 Jun 43
DV968	461	SD Biscay 13 Aug 43
DV969	10 RAAF	FTR Biscay 21 Sep 43
DV970	228, 422, 4 OTU	SOC 25 Apr 45
DV971	119	FTR 15 Dec 42
DV972	119	Ditch Bristol Channel 25 Nov 42
DV973	95	Xxd Bathurst 13 Apr 44
DV974	204, 95, 204	Xxd Bathurst 1 Oct 43
DV975	95	Xxd Bunce River 28 Nov 42
DV976	MAEE	Ditch Southend 21 Oct 47
DV977	228	FTR Biscay 12 Jul 43
DV978	246, 228, 423, 131 OTU	DBR Lough Erne Dec 44
DV979	246	Xxd Black Rock 24 Jan 43
DV980	246, 228, 423, 131 OTU	SOC 14 Jun 45
Sunderland III – Short Bros (Belfast) (20)		
DV985	461, 308 FTU, 343	Xxd Goree Island 26 Apr 44
DV986	461, 308 FTU, 343	Xxd Port Etienne 19 Sep 43
DV987	308 FTU, Flt 7E, 343	SOC 21 Jun 45
DV988	228, 422, 4 OTU	SOC 26 Mar 47
DV989	461, 131 OTU	SOC 28 Jun 45
DV990	422	FTR 24 May 44
DV991	204	DBR Jui 13 Jul 44
DV992	330, 4 OTU	Xxd Alness 6 Nov 44
DV993	10 RAAF	SD Biscay 17 Nov 43
DV994	422, 4 OTU	SOC 26 Mar 47
DW104	204	SOC 25 Oct 45
DW105	95	Xxd Port Etienne 5 Jan 44
DW106	270	FTR 18 Dec 43
DW107	95	Sank Bathurst 15 Oct 44
DW108	270	Xxd Jui 27 Sep 44
DW109	270	Scrap 10 Mar 47
DW110	228	Xxd Blue Stack Mts 31 Jan 44
DW111	270, 228, 423, 4 OTU	Scrap 10 Mar 47
DW112	423, 302 FTU, 230, Iraq com flt	SOC 29 Aug 46
DW113	10 RAAF	SOC 28 Dec 45
Sunderland III – Shorts (Rochester) (15)		
EJ131	230	FTR 27 Mar 43
EJ132	461, 230	SOC 29 May 45
EJ133	119, 461	DBR Dec 44
EJ134	461	DBR 2 Jun 43
EJ135	230, 302 FTU, 490	To Aeronavale Jun 45
EJ136	230	SOC 15 Apr 45
EJ137	246,330, 201, 4 OTU	SOC 28 Jun 45
EJ138	119, 461, 330	DBR 2 Nov 45
EJ139	246, 228	FTR Biscay 24 May 43
EJ140	230	Xxd Voi, Kenya 29 Dec 43
EJ141	230, 205	SOC 22 Jun 45
EJ142	119, 461, 4 OTU	SOC 12 Jul 45
EJ143	230	SOC 12 Mar 45
EJ144	95	Sank Bathurst 15 Oct 44
EJ145	204	Ditch Rio de Oro 17 Jul 43

Aircraft	Units	Fate
Sunderland III – Shorts (Windermere) (10)		
EJ149	4 OTU	SOC 26 Mar 47
EJ150	201	SOC 23 Mar 47
EJ151	201 228, 422	SOC 28 Jun 45
EJ152	4 OTU	(Conv Mk V)
EJ153	461, 235 OCU, 230	(Conv Mk V) SOC 1 Nov 56
EJ154	461	Xxd Pembroke Dock 13 Dec 44
EJ155	230, 4 OTU, 88, 209, 205	(Conv Mk V) scrap Oct 57
EJ156	423	Sold 13 May 47
EJ157	423	DBR Castle Archdale 12 May 45
EJ158	423	SOC 26 Mar 47
Sunderland III – Shorts (Belfast) (10)		
EJ163	302 FTU, 95, 343	To Aeronavale Jun 45
EJ164	308 FTU, 270	Ditch 3 Oct 44
EJ165	302 FTU, 490	Scrap Mar 47
EJ166		DBR Nov 43
EJ167		To RNZAF as NZ4116 4 Jun 53
EJ168	302 FTU, 343	To Aeronavale Jun 45
EJ169	302 FTU, 95	Scrap Mar 47
EJ170		Sold 13 May 47
EJ171	302 FTU	(Conv Mk V) 22 Jan 48
EJ172	302 FTU	(Conv Mk V) 2 Jun 47
Sunderland III – Blackburn (25)		
EK572	228	FTR Biscay 11 Nov 43
EK573	10 RAAF, 4 OTU	SOC 23 Mar 47
EK574	10 RAAF	Sank Plymouth 1 Jun 44
EK575	461, 228, 423, 10 RAAF	SOC 30 Nov 45
EK576	422, 4 OTU	SOC 12 Jul 45
EK577	461, 4 OTU	SOC 15 Jul 45
EK578	461	Ditch Biscay 16 Sep 43
EK579	201, 131 OTU	Sold 22 Jan 48 (LV-ANG)
EK580	204	Xxd Bathurst 9 Oct 44
EK581	423	SOC 26 Mar 47
EK582	204	Scuttled 21 Jun 45
EK583	423, 131 OTU	SOC Mar 47
EK584	270	SOC 2 Jun 45
EK585	270	Sank Apapa 7 Jun 44
EK586	10 RAAF	SOC 31 Dec 45
EK587	95	To Aeronavale Jun 45
EK588	270	Xxd 8 Jan 44
EK589	270	SOC 9 Jul 45
EK590	201, 461, 4 OTU	SOC 26 Mar 47
EK591	422, 4 OTU	SOC 2 Nov 45
EK592	270	(Conv GR.5) scrap Mar 47
EK593	302 FTU, 270	Scrap Mar 47
EK594	201, 422, 10 RAAF	SOC 12 Aug 45
EK595	302 FTU, 201, 422, Iraq CF	DBR Basra 5 Mar 46
EK596	302 FTU	Scrap Mar 47
Sunderland III – Shorts (Rochester) (50)		
JM659	230	Sank Dibrugarh 4 Jul 44
JM660		To BOAC 8 Jan 43 (G-AGER)
JM661		To BOAC 15 Jan 43 (G-AGES)
JM662		To BOAC 16 Jan 43 (G-AGET)
JM663		To BOAC 21 Jan 43 (G-AGEU)
JM664		To BOAC 29 Jan 43 (G-AGEV)
JM665		To BOAC 29 Jan 43 (G-AGEW)

Aircraft	Units	Fate
JM666	330, 201, 423, 131 OTU	Scrap 10 Mar 47
JM667	330, 302 FTU, 209, 205	(Conv Mk V) SOC 8 Oct 54
JM668	4 OTU	Scrap Mar 47
JM669	204	Ditch 14 Apr 43
JM670	95, 343	SOC 21 Jun 45
JM671	95	SOC 21 Jun 45
JM672	204	Xxd Jui 28 Aug 44
JM673	230	FTR 28 Nov 44
JM674	204	SOC 21 Jun 45
JM675	461	Xxd Biscay 28 May 43
JM676	119, 461	SD Biscay 29 Nov 43
JM677	95	SOC 21 Jun 45
JM678	228, 461, 228, 10 RAAF	Sank Mount Batten 19 Jun 44
JM679	228, 422, 4 OTU	SOC 6 Jul 45
JM680	204	Sank Half Die 31 May 43
JM681	MAEE	(Conv Mk V) Sold 8 May 47 (G-AJMZ)
JM682	204	Scrap Feb 47
JM683	461, 4 OTU	SOC 12 Jul 45
JM684	10 RAAF	SOC 12 Jul 45
JM685	461, 228, 10 RAAF	SOC 7 Jul 45
JM686	461, 4 OTU	SOC 28 Jun 45
JM687	204	FTR 18 Jul 43
JM688	308 FTU, 343	SOC 21 Jun 45
JM689	308 FTU, 343	To Aeronavale Jun 45
JM704	308 FTU, 343	Xxd Dakar 5 Feb 44
JM705	308 FTU, Flot 7FE	Xxd Dakar 27 Oct 43
JM706	308 FTU, Flot 7FE, 343	SOC 21 Jun 45
JM707	461	SD Biscay 30 Aug 43
JM708	228	Ditch Biscay 17 Jan 44
JM709	228	FTR Biscay 6 Jan 44
JM710	204	Xxd Bathurst 22 Sep 43
JM711	308 FTU, 230, Iraq CF	SOC 29 Aug 46
JM712	422	Ditch 17 Nov 43
JM713	MAEE, 4 OTU	Sold Aug 47 (LN-LMK)
JM714	MAEE, 302 FTU	(Conv Mk V) Sold 22 Jan 48
JM715		(Conv Mk V) sold 30 Apr 47 (ZK-AMH)
JM716		To BOAC 7 Mar 45 (G-AHEO)
JM717	302 FTU, 490	(Conv Mk V) SOC 21 Jun 45
JM718	4 OTU, 230, 235 OCU	(Conv Mk V) Scrap Oct 57
JM719	302 FTU	(Conv Mk V) Sold May 47 (G-AKCO)
JM720	228, 302 FTU	(Conv Mk V) Sold Oct 46 (LN-IAW)
JM721	10 RAAF	Sank Loch Ryan 17 Nov 44
JM722		To BOAC 21 Aug 43 (G-AGHV)
Sunderland III – Shorts (Rochester) (75)		
ML725		To BOAC 27 Aug 43 (G-AGHW)
ML726		To BOAC 2 Sep 43 (G-AGHX)
ML727		To BOAC 3 Sep 43 (G-AGHZ)
ML728		To BOAC 8 Sep 43 (G-AGIA)
ML729		To BOAC 15 Sep 43 (G-AGIB)
ML730		To RAAF as A26-1 Nov 43
ML731		To RAAF as A26-2 Nov 43
ML732		To RAAF as A26-3 Nov 43
ML733		To RAAF as A26-4 Nov 43
ML734		To RAAF as A26-5 Nov 43
ML735	MAEE, 461	FTR Bergen 1 Oct 44
ML736	4 OTU	SOC 26 Mar 47
ML737	4 OTU	DBR Alness 3 Feb 45
ML738	4 OTU	Sank Alness 3 Jan 45

ML742 in service with 201 Squadron at Lough Erne. This aircraft was scrapped on 26 March 1947.
Peter Green Collection

Aircraft	Units	Fate
ML739	201, 461, 4 OTU, 302 FTU	(Conv GR.V) To Aeronavale 19 Feb 52
ML740	461	Ditch Biscay 23 Mar 44
ML741	461, 4 OTU, CCIS, 4 OTU, 302 FTU	Scrap Apr 50 (Conv GR.V)
ML742	201	Scrap 26 Mar 47
ML743	461, 201	Xxd Killybegs 14 Mar 45
ML744	461, 422	SOC 26 Mar 47
ML745	228, 88, 205	(Conv GR.V) SOC 24 May 57
ML746	461, 423	SOC 26 Mar 47
ML747	461, 4 OTU	(Conv GR.V) SOC 26 Aug 46
ML748	461	Sank Scillies 11 Jun 44
ML749	201, 228, 4 OTU	Scrap 10 Mar 47
ML750	MAEE, 422	To Aeronavale 26 Jul 47
ML751		To BOAC 13 Jan 44 (G-AGJJ)
ML752		To BOAC 21 Jan 44 (G-AGJK)
ML753		To BOAC 23 Jan 44 (G-AGJL)
ML754		To BOAC 2 Feb 44 (G-AGJM)
ML755		To BOAC 3 Feb 44 (G-AGJN)
ML756		To BOAC 3 Feb 44 (G-AGJO)
ML757	461, 4 OTU, 302 FTU	(Conv GR.V) To Aeronavale 19 Jun 51
ML758	461, 422, 330	Scrap 26 Mar 47
ML759	201, 422	Scrap 26 Mar 47
ML760	201	SD N. Atlantic 12 Jun 44
ML761		(Conv GR.V) Sold 18 Mar 46
ML762	228	FTR Biscay 10 Jun 44
ML763	228, ASWDU, 230	(Conv GR.V) SOC 30 Sep 57
ML764	201	(Conv GR.V) To Aeronavale 26 Jul 51
ML765	MAEE	SOC 6 Nov 47
ML766	228	DBR Pembroke Dock 14 Nov 44
ML767	228, 4 OTU	SOC 26 Mar 47
ML768	201	(Conv GR.V) Scrap 25 Aug 55
ML769	201, 228, 422	Sold 25 Aug 47
ML770	228	Sank Scillies 21 Feb 45
ML771	461	Scrap 10 Mar 47
ML772	201, 88	(Conv GR.V) SOC 30 Jun 55
ML773	422	SOC 26 Mar 47
ML774	228, 461	Sank Angle Bay 18 Jan 45
ML777	422, 423	Scrap 21 Jan 47
ML778	422, 461, 201, 4 OTU	(Conv GR.V) To Aeronavale 12 May 51

Aircraft	Units	Fate
ML779	4 OTU, 330	(Conv GR.V) To Aeronavale 27 Apr 51
ML780	330, 4 OTU, 235 OCU	Scrap 11 Aug 48
ML781	422, 461, 4 OTU, CCIS	To Aeronavale 14 Apr 51
ML782	201, 228	Sank Mount Batten 11 Dec 44
ML783	201, 423, 201	Sold 15 Apr 46 (G-AHZA)
ML784	423, 201	Sold 12 Apr 46 (G-AHIZ)
ML785	ASWDU	Sold 22 Aug 47
ML786		To BOAC 12 Jul 44 (G-AGKV)
ML787		To BOAC 25 Jul 44 (G-AGKW)
ML788		To BOAC 28 Jul 44 (G-AGKX)
ML789		To BOAC 28 Jul 44 (G-AGKY)
ML790		To BOAC 3 Aug 44 (G-AGKZ)
ML791		To BOAC 11 Aug 44 (G-AGLA)
ML792	302 FTU	To RNZAF as NZ4101 18 Oct 44
ML793	302 FTU	To RNZAF as NZ4102 23 Oct 44
ML794	302 FTU	To RNZAF as NZ4103 18 Oct 44
ML795	302 FTU	To RNZAF as NZ4104 27 Oct 44
ML796	4 OTU	To Aeronavale 4 Aug 51 Preserved
ML797	308 FTU, 230, 205	SOC 30 Jun 59
ML798		To SAAF as 1709 31 May 45
ML799	302 FTU	To Aeronavale 4 Jun 51
ML800	302 FTU, 230	To Aeronavale 28 Jul 57
ML801	4 OTU, 302 FTU	Scrap 10 Mar 47
Sunderland III – Short Bros (Belfast) (25)		
ML807		(Conv GR.V) Sold 19 Sep 46 (LN-IAU)
ML808	131 OTU	Sold 22 Aug 47
ML809	302 FTU	Sold 7 Sep 46 (LN-IAV)
ML810	302 FTU, 490	SOC 21 Jun 45
ML811	302 FTU	Xxd Tornardo Lakka 3 Jun 44
ML812	228, 201, 4 OTU, 302 FTU	(Conv GR.V) SOC 30 Jun 55
ML813	10 RAAF, 201	Scrap 26 Mar 47
ML814	201, 422, 330	(Conv GR.V) To RNZAF as NZ4108 26 May 53
ML815	228, 235 OCU	(Conv GR.V) SOC 20 Sep 57
ML816	422, 4 OTU	(Conv GR.V) To Aeronavale 10 Aug 51
ML817	MAEE, 201, 423, 330, 235 OCU, 230	SOC 16 Oct 57
ML818	330	(Conv GR.V) Sold 30 May 46 (G-AHZE)
ML819	330	(Conv GR.V) To Aeronavale 30 Sep 51
ML820	4 OTU	(Conv GR.V) To Aeronavale 23 Nov 51
ML821	422	(Conv GR.V) To Aeronavale 26 Sep 51
ML822	10 RAAF	SOC 12 Jul 45
ML823	423	Xxd Donegal Bay 6 Sep 44
ML824	201, 330	To Aeronavale 26 Oct 51 Preserved
ML825	423	Scrap 26 Mar 47
ML826	4 OTU	To 6100M 31 Jul 46
ML827	461, 330	(Conv GR.V) Ditch 12 May 45
ML828	10 RAAF	(Conv GR.V) (G-AHZG)
ML829	10 RAAF	Sank Mount Batten 9 Feb 45
ML830	10 RAAF	Scrap 10 Mar 47
ML831	10 RAAF, 461	DBR Angle Bay 18 Jan 45
Sunderland III – Blackburn (50)		
ML835	302 FTU, 490	To Aeronavale 21 Jun 45
ML836	422, 131 OUT	SOC 31 May 45
ML837	302 FTU, 95	SOC 21 Jun 45
ML838		To BOAC 5 Apr 46 (G-AHYY)
ML839	10 RAAF	Sank Mount Batten 12 Oct 44
ML840		(Conv GR.V) Sold 2 Jun 47 (G-AKCR)

Aircraft	Units	Fate
ML841	4 OTU, 302 FTU, 343	To Aeronavale 21 Jun 45
ML842	131 OTU	SOC 6 Sep 45
ML843		Sold May 46 (G-AHRE)
ML844	270	SOC 21 Jun 45
ML845	302 FTU	Ditch Canaries 23 Jul 44
ML846	302 FTU, 230	SOC 13 Sep 45
ML847	302 FTU, 95	SOC 21 Jun 45
ML848	10 RAAF	SOC 26 Mar 47
ML849	302 FTU, 270	Scrap 10 Mar 47
ML850	302 FTU, 490	Scrap 10 Mar 47
ML851	302 FTU, 343	To Aeronavale 21 Jun 45
ML852	302 FTU, 490	Ditch Cape St Mary 14 Jul 44
ML853	302 FTU, 270	Scrap 10 Mar 47
ML854	302 FTU, 204, 343	To Aeronavale 21 Jun 45
ML855	302 FTU	Xxd St Louis 17 Jul 44
ML856	10 RAAF	SOC 30 Nov 45
ML857	302 FTU, 490, 270	SOC 21 Jun 45
ML858	302 FTU	Xxd St Kilda 8 Jun 44
ML859	302 FTU, 490	SOC 21 Jun 45
ML860	302 FTU	Ditch 23 Jul 44
ML861	302 FTU, 230	SOC 28 Jun 45
ML862	302 FTU. 490, 204	SOC 21 Jun 45
ML863	302 FTU, 490	SOC 21 Jun 45
ML864	302 FTU, 490	SOC 16 May 45
ML865	302 FTU, 230, SF Koggala	SOC 28 Jun 45
ML866	4 OTU	(Conv GR.V) To Aeronavale 19 Nov 51
ML867	302 FTU, 270	SOC 21 Jun 45
ML868	302 FTU, 230	SOC 31 Jan 46
ML869	302 FTU, 490	Scrap 10 Mar 47
ML870	302 FTU, 343	SOC 21 Jun 45
ML871	302 FTU, 343	To Aeronavale 21 Jun 45
ML872	302 FTU, 204	(Conv GR.V) To Aeronavale 14 Dec 51
ML873	4 OTU	(Conv GR.V) Scrap 25 Aug 55
ML874	302 FTU, 270	To Aeronavale 21 Jun 45
ML875	201, 4 OTU	(Conv GR.V) SOC 6 Nov 47
ML876	201	Sold 29 Mar 46 (G-AGWX)
ML877	228	(Conv GR.V) To Aeronavale 12 Jan 52
ML878	228, 330, 302 FTU	(Conv GR.V) SOC 25 Sep 47
ML879	228, 461, 422	Sold 17 Nov 47
ML880	228	FTR Biscay 12 Jun 44
ML881	201, 209	(Conv GR.V) Scrap 20 Sep 57
ML882	201, 4 OTU, 88, 209	(Conv GR.V) Scrap 19 Oct 56
ML883	422, 423	Sank Calshot 17 Dec 44
ML884	422	SOC 26 Mar 47
Sunderland III – Blackburn (25)		
NJ170	422, 330	(Conv GR.5) To Aeronavale 25 May 51
NJ171	228	(Conv GR.5) To BOAC 8 May 46 (G-AHZB)
NJ172	422, 330	(Conv GR.5) SOC 30 Sep 57
NJ173	422	SOC 26 Mar 47
NJ174	422	SOC 26 Mar 47
NJ175	422	Xxd Belleek 12 Aug 44
NJ176	422, 88	(Conv GR.5) Xxd Seletar 20 Nov 49
NJ177	330, 209	(Conv GR.5) SOC 19 Aug 54
NJ178	330	SOC 26 Mar 47
NJ179	330	(Conv GR.5) Sold (ZK-AME)
NJ180	330, FBTS	(Conv GR.5) DBR Pembroke Dock 23 Feb 54

NJ176 crashed during a night take-off from Seletar on 20 November 1949. Ken Delve Collection

Aircraft	Units	Fate
NJ181	330	FTR 4 Oct 44
NJ182	423, ASWDU	(Conv GR.5) To Aeronavale 8 Jan 52
NJ183	423	Xxd Irvinestown 11 Feb 45
NJ184	423	SOC 3 Aug 45
NJ185	423	SOC 26 Mar 47
NJ186	423	Xxd Jurby 20 May 45
NJ187	423	(Conv GR.5) SOC 30 Apr 57
NJ188	330	(Conv GR.5) Sold 20 May 46 (G-AHZF)
NJ189	422	SOC 26 Mar 47
NJ190	201, 330, 4 OTU	(Conv GR.5) To Aeronavale 19 Feb 52
NJ191	228, 4 OTU, 209, 205	(Conv GR.5) SOC 30 Nov 56
NJ192	228, 201, 4 OTU	(Conv GR.5) SOC 20 Feb 46
NJ193	461, 10 RAAF, 201, 205	(Conv GR.5) SOC 28 Feb 57
NJ194	201	(Conv GR.5) Scrap 31 Aug 53
Sunderland V – Short Bros (30)		
PP103	302 FTU, 209	Xxd Seletar 27 Mar 46
PP104	302 FTU	To SAAF 31 May 45
PP105	302 FTU, 209	SOC 10 Mar 47
PP106	302 FTU, 209	SOC 10 Mar 47
PP107	302 FTU, 209, 205	Xxd Hwalien 28 Jan 51
PP108	302 FTU, 209	Scrap 10 Mar 47
PP109	MAEE, 302 FTU	To SAAF 13 Jun 45
PP110		To RNZAF as NZ4105 15 Jan 45
PP111	4 OTU	SOC Dec 45
PP112	4 OTU, 201, 209	SOC 30 Jun 58
PP113	4 OTU, 461, 10 RAAF, 201	Xxd Inistrahull Island 5 Jul 47
PP114	461, 10 RAAF, 230, 201, 209, 88	SOC 31 Aug 55
PP115	461, 10 RAAF, 230, 201, 230, 201	SOC 30 Jun 55
PP116	461	DBR Pembroke Dock 16 May 45
PP117	228, 201, 230, 201	SOC 4 Oct 57
PP118	228, 201, 230, 235 OCU	Sank Calshot 3 Feb 50
PP119	461, 10 RAAF, 201	SOC 6 Nov 47
PP120	228, 201	SOC 27 Mar 47
PP121	228, 201, 57 MU	SOC 23 Apr 46
PP122	461, 10 RAAF, 201, 230, ASWDU, 201	Sank St Peter Port 15 Sep 54
PP123	302 FTU, 205, 57 MU	Sank Wig Bay 2 Dec 48
PP124	302 FTU, 205	To RNZAF as NZ4113

Aircraft	Units	Fate
PP125	302 FTU	To SAAF 9 Sep 45
PP126	302 FTU, 240	SOC 11 Aug 47
PP127	205	SOC 1 Jun 57
PP128	302 FTU, 205	SOC 10 Mar 47
PP129	302 FTU, 205	To RNZAF as NZ4110 1 Jul 53
PP130	302 FTU, 240, 209, 235 OCU	SOC 30 Jun 55
PP131	302 FTU, 240	SOC 1 Sep 53
PP132	302 FTU, 209	DBR Kai Tak 21 Apr 46
Sunderland III – Short Bros (10)		
PP135	10 RAAF	SOC 12 Jul 45
PP136	228	DBR Angle Bay 18 Jan 45
PP137	4 OTU, 205	(Conv Mk V) SOC 11 Feb 57
PP138	10 RAAF	SOC 10 Mar 47
PP139	10 RAAF	SOC 10 Mar 47
PP140	330	Xxd 5 Apr 45
PP141	4 OTU, 235 OCU	(Conv Mk V) SOC 21 Aug 55
PP142	10 RAAF	To BOAC as G-AHER 22 Jun 49
PP143		(Conv Mk V) To RNZAF as NZ4119 19 Jan 45
PP144	228, 201, 205	(Conv Mk V) SOC 30 Jun 55
Sunderland V – Blackburn (19)		
PP145	302 FTU, 230	DBR 29 Apr 46 (to 6103M)
PP146	302 FTU, 230	SOC 6 Nov 47
PP147	302 FTU, 230	SOC 16 Oct 57
PP148	302 FTU, 209, 205, 88	Xxd Iwakuni 25 Mar 53
PP149	302 FTU, 230	SOC 21 May 54
PP150	302 FTU, 209	SOC 26 Sep 46
PP151	302 FTU, 209, MAEE, Manf	SOC 31 Oct 56
PP152	302 FTU	To SAAF 26 Apr 45
PP154	302 FTU, 205, 209	SOC 30 Sep 57
PP155	302 FTU, 230, 88, 230	Xxd Transgigvaag 23 Oct 54
PP156	302 FTU	To SAAF 31 May 45
PP157	302 FTU, 230	DBR Jesselton 19 Jan 46
PP158	302 FTU, 230	DBR Kuantan 3 Oct 45
PP159	302 FTU, 209	SOC 6 Nov 47
PP160	4 OTU	SOC 24 Apr 47
PP161	4 OTU, 230	SOC 6 Nov 47
PP162	461, 10 RAAF, 201, MAEE	Sold 18 Jun 53
PP163	228, 201, 235 OCU	SOC 16 Oct 57
PP164	228, 201, 230, 209	DBR Yokohama 19 Jul 50
Sunderland V – Short Bros (10)		
RN264	302 FTU, 209	DBR Sama Bay 18 Jul 46
RN265	302 FTU, 209	DBR Seletar 23 Jan 46
RN266	302 FTU, 209, 201, 235 OCU, FBTS	SOC 30 Sep 57
RN267	330	SOC 6 Nov 47
RN268	302 FTU, 230	SOC 4 May 53
RN269	302 FTU, 230, 201, 205	DBR Sangley Point 28 Jun 43
RN270	201, 230, 205	SOC 325 Sep 58
RN271	235 OCU, 201	SOC 30 Sep 57
RN272	201, 4 OTU, 235 OCU	To 6534M Apr 48
RN273	201, 205	SOC 13 Mar 57
Sunderland V – Blackburn (30)		
RN277	228, 201, 88, 209	DBR Iwakuni 14 Oct 51
RN278	228, 201, 230, 205	SOC 27 Aug 56
RN279	461, 302 FTU	To SAAF 13 Jun 45
RN280	461, 302 FTU, SF Koggala, 205	To RNZAF as NZ4106 21 Dec 53
RN281	302 FTU	To SAAF 1 Jun 45

RN294 with 205 Squadron carrying out a VJ-Day flypast over Colombo. Andy Thomas Collection

Aircraft	Units	Fate
RN282	461, 10 RAAF, 201, 88, 209, 205	SOC 13 May 58
RN283	228	Xxd 27 Apr 45
RN284	201, 235 OCU, 201	To Aeronavale 18 Dec 57
RN285	228, 201, 4 OTU	Scrap 4 May 53
RN286	4 OTU	To RNZAF as NZ4117 3 Oct 53
RN287	4 OTU, 302 FTU	SOC 31 Jul 46
RN288	302 FTU, 205, 209, 235 OCU, 201	Xxd Eastbourne 4 Jun 55
RN289	302 FTU	SOC 6 Nov 47
RN290	302 FTU, 205, 230	Xxd Angle Bay 30 Jan 54
RN291	302 FTU, 240, 88	To RNZAF as NZ1120 15 Feb 54
RN292	302 FTU, 240	DBR Wig Bay 14 Oct 48
RN293	302 FTU, 205, 209, 88, 209, 205	SOC 17 Aug 56
RN294	302 FTU, 205	Sank Wig Bay 20 Dec 51
RN295	302 FTU	To SAAF 31 Jul 45
RN296	302 FTU	To SAAF 27 Jul 45
RN297	302 FTU, 240, MAEE	Sold 18 Jun 53
RN298	302 FTU, 240, 209	Scrap 5 Jul 50
RN299	302 FTU, 230, 201	SOC 4 Oct 57
RN300	10 RAAF, 201, 205, 209, 205	SOC 30 Apr 57
RN301	302 FTU, 205	SOC 30 Jun 51
RN302	302 FTU, SF Koggala, 235 OCU, 209, 88	Xxd Tsushima Island 27 Dec 53
RN303	302 FTU, 230, 209, FBTS, 230, 205	SOC 24 Jan 59
RN304	302 FTU, 201, 230, FBTS	SOC 20 Sep 57
RN305	302 FTU	To SAAF 12 Oct 45
RN306	302 FTU, 205	To RNZAF as NZ4118 15 Dec 53
Sunderland V – Short Bros (27)		
SZ559	302 FTU, 209	DBR Sana Bay 18 Jul 46
SZ560	302 FTU, 205, 209, FBTS, 230	SOC 15 Oct 57
SZ561	302 FTU, TRE, 209	To RNZAF as NZ4114 18 Oct 53
SZ562	302 FTU, 209	SOC 15 Aug 46
SZ563	302 FTU	SOC 25 Apr 46
SZ564	302 FTU, 240	DBR Kai Tak 26 Sep 46
SZ565	302 FTU, 209, 201, 235 OCU	Xxd Hillshead 16 Nov 51
SZ566	302 FTU, 205, 209, 205, 88	SOC 15 Oct 57
SZ568	302 FTU, 4 OTU, 235 OCU	SOC 19 Oct 56
SZ569	302 FTU, 4 OTU, 235 OCU, 205	DBR Trincomalee 3 Oct 50
SZ570	302 FTU, 88	DBR Kalafrana 28 Feb 48
SZ571	302 FTU, 4 OTU, 201, 209, 88, FBTS, 201	To Aeronavale 14 Nov 57
SZ572	230, 88, 205	SOC 19 Jul 57
SZ573	BOAC, 230, 209	Sank Seletar 26 Mar 50

SZ559 was damaged beyond repair at Sana Bay 18 July 1946. Ken Delve Collection

Aircraft	Units	Fate
SZ574	201	DBR Lough Erne 31 May 48
SZ575	302 FTU, 4 OTU, 235 OCU, 230, 201	SOC 15 Oct 57
SZ576	235 OCU, 201	To Aeronavale 4 Jul 57
SZ577	230, 88, 209, 205	SOC 21 May 57
SZ578	201, 209, 205, 88, 205	SOC 16 Oct 57
SZ579		Wrecked pre delivery
SZ580	235 OCU	SOC 30 Sep 57
SZ581	230	Sank Wig Bay 2 Nov 55
SZ582	230	SOC 30 Sep 57
SZ583		SOC 18 Aug 59
SZ584	BOAC	To RNZAF as NZ4115 4 Sep 53
SZ598	201	Xxd Beja 16 Feb 51
SZ599	MAEE, 209, 88, 209	SOC 21 Jun 51
SZ600–SZ611		Cancelled
Sunderland V – Short Bros (1)		
TX293	MAEE, 57 MU	DBR Wig Bay 9 Feb 48
TX294–TX296	Cancelled	
Sunderland GR.5 – Blackburn (10)		
VB880	302 FTU, 88	To RNZAF as NZ4111 6 Jul 53
VB881	201	To RNZAF as NZ4112 10 Aug 53
VB882	302 FTU, 230, 209	SOC 5 Jun 50
VB883	302 FTU, 88	To RNZAF as NZ4107 3 Nov 53
VB884	302 FTU, 209	SOC 26 Aug 52
VB885	302 FTU	Xxd Calshot 13 Feb 46
VB886	4 OTU	Sank 16 Mar 47

Aircraft	Units	Fate
VB887	230, 88	SOC 18 Jul 54
VB888	302 FTU, 4 OTU, 235 OCU, 88, 209	SOC 16 Aug 56
VB889	201	SOC 8 Aug 56

Civil Serials – BOAC

Sunderland III – acquired 1942

G-AGER	ex-JM660	Scrap Jul 56
G-AGES	ex-JM661	Xxd Brandon Head 28 Jul 43
G-AGET	ex-JM662	DBR River Hooghly 15 Feb 43
G-AGEU	ex-JM663	Scrap Aug 53
G-AGEV	ex-JM664	DBR Poole Harbour 4 Mar 46
G-AGEW	ex-JM665	Sank Sourabaya 5 Sep 48

Sunderland III – acquired 1943

G-AGHV	ex-JM772	Sank Rod-el-Faraq 10 Mar 46
G-AGHW	ex-ML725	Xxd Isle of Wight 19 Nov 47
G-AGHX	ex-ML726	Scrap Oct 48
G-AGHZ	ex-ML727	Scrap Jan 52
G-AGIA	ex-ML728	Scrap Jul 52
G-AGIB	ex-ML729	Xxd Sollum 5/6 Nov 43
G-AGJJ	ex-ML751	Scrap Jan 52
G-AGJK	ex-ML752	Scrap Jan 52
G-AGJL	ex-ML753	Scrap Jan 52
G-AGJM	ex-ML754	Scrap Jan 52
G-AGJN	ex-ML755	DBR Madeira 21 Jan 53
G-AGJO	ex-ML756	DBR Hythe 21 Feb 49

Sunderland III – acquired 1944

G-AGKV	ex-ML786	Scrap May 51
G-AGKW	ex-ML787	Scrap May 51
G-AGKX	ex-ML788	Scrap 53
G-AGKY	ex-ML789	DBR Calshot 21 Jan 53
G-AGKZ	ex-ML790	Scrap May 49
G-AGLA	ex-ML791	Scrap Sep 49

Sunderland III

G-AHEO	ex-JM716	Scrap Nov 49
G-AHEP	ex-DD860	Wfu Sep 52
G-AHER	ex-PP142	Scrap Jan 52

Sandringham 2 (Conv from Sunderland)

G-AHRE	ex-ML843	Became LV-ACT

Sandringham 3 (Conv from Sunderland)

G-AGWW	ex-EJ156	Became CX-AFA
G-AGWX	ex-ML876	Became CX-AKF

Sandringham 5 (Conv from Sunderland)

G-AHYY	ex-ML838	Scrap Mar 59
G-AHYZ	ex-ML784	DBR Belfast Jan 47
G-AHZA	ex-ML783	Scrap Mar 59
G-AHZB	ex-NJ171	Xxd Bahrain 23 Aug 47
G-AHZC	ex-NJ253	Scrap Mar 59
G-AHZD	ex-NJ257	Became VH-EBV
G-AHZE	ex-ML818	
G-AHZF	ex-NJ188	Became VH-EBY
G-AHZH	ex-ML828	Became VH-EBZ
G-AJMZ	ex-JM681	Scrap Mar 59

Sandringham 7 (Conv from Sunderland)

G-AKCO	ex-JM719	Became VH-APG
G-AKCP	ex-EJ172	Became CX-ANI
G-AKCR	ex-ML840	Became CX-ANA

Sunderland Squadrons

88 Sqn
Marks: GR.5 Sep 46–Oct 54
Bases: Kai Tak, Seletar

95 Sqn *code:* SE
Marks: I Jan 41–Jan 44
III Jul 42–Jun 45
Bases: Pembroke Dock, Freetown, Jui, Bathurst

119 Sqn
Marks: II Sep 42–Apr 43
III Sep 42–Apr 43
Bases: Pembroke Dock

201 Sqn *Code:* NS, ZM
Marks: I Apr 40–Jan 42
II May 41–Mar 44
III Jan 42–Jun 45
V Feb 45–Feb 57
Bases: Invergordon, Sullom Voe, Castle Archdale, Pembroke Dock, Calshot

202 Sqn *code:* AX, TQ
Marks: I/II Dec 41–Sep 42
III Mar 42–Sep 42
Bases: Gibraltar

204 Sqn *code:* KG, RF
Marks: I Jun 39–Sep 43
II Jun 41–Mar 43
III Oct 42–Jun 45
V Apr 45–Jun 45
Bases: Mount Batten, Sullom Voe, Reykjavik, Gibraltar, Bathurst, Half Die, Jui

88 Squadron, Kai Tak, February 1947. Gerry Paine

209 Sqn
Marks: V Feb 45–Dec 54
Bases: Kipevu, Koggala, Kai Tak, Seletar, Iwakuni

210 Sqn *code:* DA, VG
Marks: I Jun 38–Apr 41
Bases: Pembroke Dock, Tayport, Invergordon, Sullom Voe, Oban

228 Sqn *code:* DQ, UE
Marks: I Nov 38–Aug 42
II Mar 42–Nov 43
III Mar 42–Feb 45
V Feb 45–Jun 45
Bases: Pembroke Dock, Alexandria, Kalafrana, Aboukir, Half Die, Stranraer, Oban, Castle Archdale

230 Sqn *code:* DX, NM, FV, 4X
Marks: I/II Jun 38–Jan 43
III Apr 42–Mar 45
V Jan 45–Feb 57
Bases: Seletar, Penang, Trincomalee, Colombo, Koggala, Alexandria, Scaramanga, Aboukir, Kasfareet, Dar es Salaam, Akyab, Rangoon, Redhills Lake, Castle Archdale, Calshot, Pembroke Dock

(Below) **August 1943 and 204 Squadron pose for the camera.** Ken Delve Collection
(Bottom) **230 Squadron, Seletar *c*.1945, OC Wg Cdr Ted Hawkins.** Arthur Banks

461 Squadron officers, February 1944. Ken Delve Collection

490 Squadron pose in 1945; this New Zealand unit disbanded on 1 August 1945. Peter Green Collection

240 Sqn
Marks: V Jul 45–Mar 46
Bases: Redhills Lake, Koggala

246 Sqn
Marks: I/II Oct 42–Apr 43
Bases: Bowmore

259 Sqn
Marks: V Mar 45–Apr 45
Base: Dar es Salaam

270 Sqn
Marks: III Dec 43–Jun 45
Base: Apapa

330 Sqn *code:* WH
Marks: III Feb 43–May 45
V May 45–Nov 45
Bases: Oban, Sullom Voe, Stavanger

422 Sqn *code:* DG
Marks: III Nov 42–Jun 45
Bases: Oban, Bowmore, Castle Archdale, Pembroke Dock

423 Sqn *code:* AB, XI
Marks: II Jul 42–Sep 44
III Dec 42–May 45
Bases: Oban, Lough Erne, Castle Archdale, Pembroke Dock

461 Sqn *code:* UT
Marks: II Apr 42–May 43
III Aug 42–May 45
V Mar 45–Jun 45
Bases: Mount Batten, Hamworthy, Pembroke Dock, Sullom Voe

10 RAAF *code:* RB
Marks: I Sep 39–Jul 42
II Jun 41–Dec 43
III Jan 42–Jun 45
V May 44–Jun 45
Bases: Pembroke Dock, Mount Batten, Oban

Other Units

Of the following, only 4 OTU, 131 OTU and 302 FTU (Ferry Training Unit) had significant numbers of Sunderlands. The Far East Flying Boat Wing (FEFBW) was an overall name for the squadrons operating in that theatre post-war. A number of other units, such as the Iraq Communication Flight, had Sunderlands on strength from time to time.

4 (C) OTU

10 OTU

Far East Flying Boat Wing (FEFBW)

ASWDU *code:* DP

131 OTU *code:* CA, DC

Coastal Command Flying Instructors School (CCFIS) *code:* EJ

302 FTU

308 FTU

MAEE

CCDU

Sunderland Bases
(Operational Units Only)

ABOUKIR, EGYPT

AKYAB, BURMA

ALEXANDRIA, EGYPT

ALNESS/INVERGORDON

The base was operational from just before the outbreak of war, and Sunderlands commenced detachments from the latter part of 1939. Operational use by various squadrons continued for the first part of the war, but the major connection with Sunderlands came with the arrival of No. 4 (C) OTU from Stranraer in June 1941. This unit trained large numbers of aircrew and as such played a vital role in the Sunderland story. Invergordon became RAF Alness on 10 February 1943, although according to most sources this was accompanied by a move by one mile of the alighting area – *see also* OC 4 OTU's comments regarding the move. The OTU left for Pembroke Dock on 15 August 1946, although Sunderlands continued to visit Alness for some years.

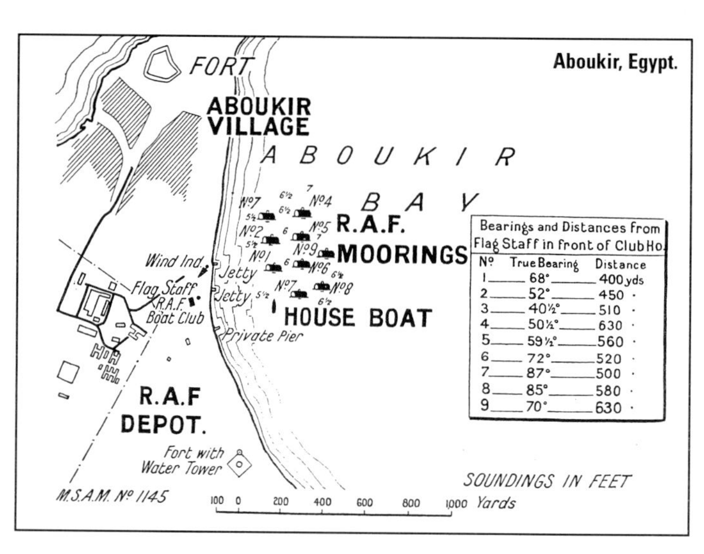

Aboukir, Egypt.

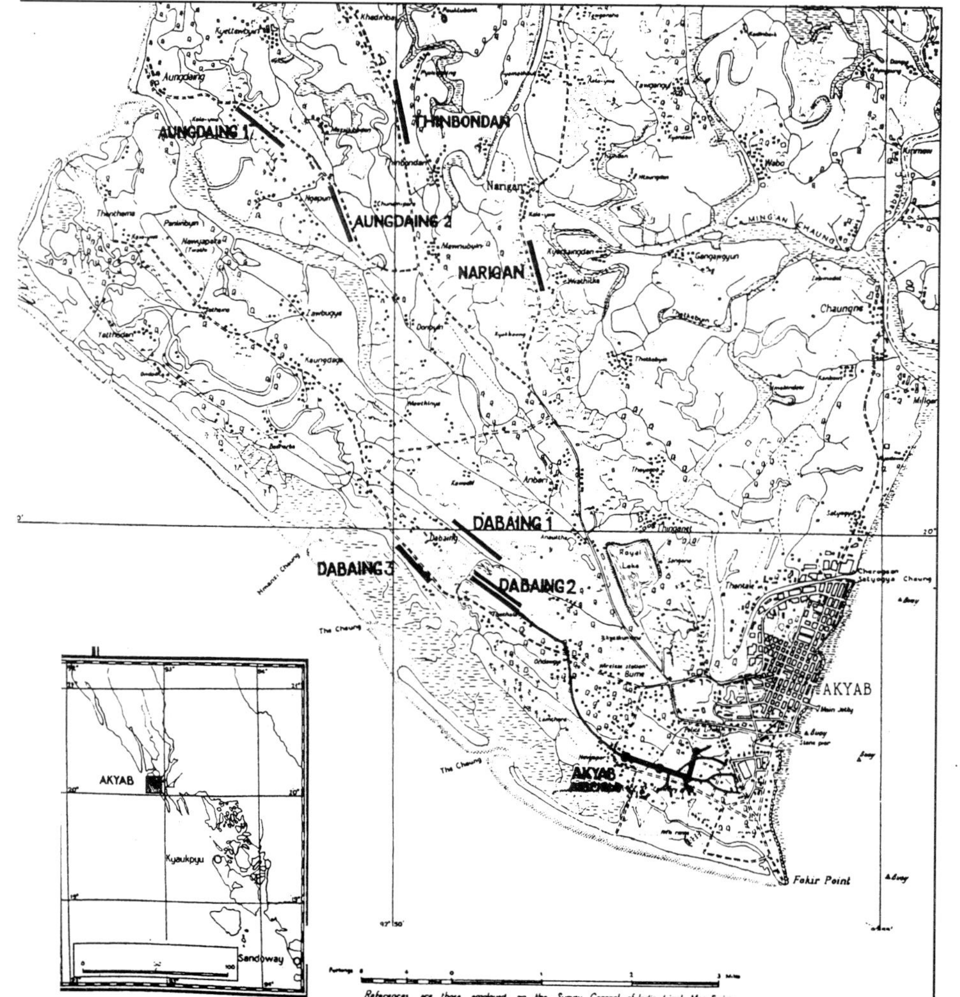

APAPA, NIGERIA

BATHURST

'Had quite a cool, agreeable climate. There were proper BOAC staff and everything ran very smoothly.' (Howard Fry)

BOWMORE

In use from late 1940, Bowmore was little more than a prepared anchorage with limited facilities and no slipway. Nevertheless, Sunderlands operated from here, as did BOAC G-Class boats, a number of the latter adopting operational duties with 119 Squadron. From September 1942 the base was home to 246 Squadron for about a year and then subsequently by a detachment of 422 Squadron, the latter being withdrawn towards the end of 1943. Although Sunderlands used Bowmore from time to time no units were stationed here and the base closed in July 1945.

CALSHOT

This Hampshire flying boat base was one of the most significant, and famous, in the UK, its origins lying in March 1913 as a seaplane base. It remained an important base in the inter-war period and was a logical early site for a Sunderland unit.

However, during the war it was not used by operational units, being home to the Flying Boat Servicing Unit (FBS). The Squadrons returned in 1946, 201

Akyab Seaplane Station, Burma.

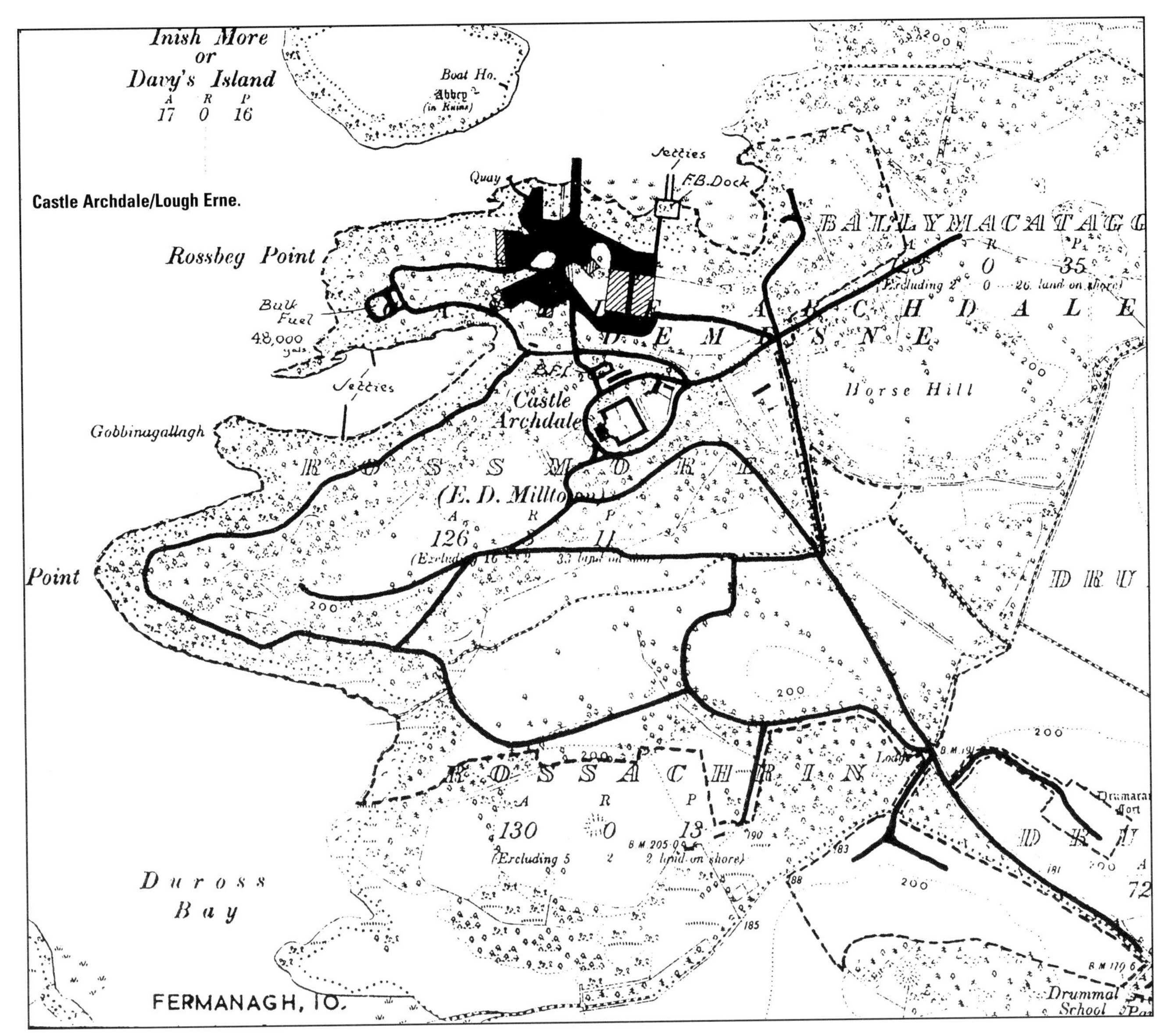

Castle Archdale/Lough Erne.

Squadron and 230 Squadron taking up residence for three years, being joined in 1947 by 4 (C) OTU (re-designated as 235 OCU). The latter unit remained in place until October 1953.

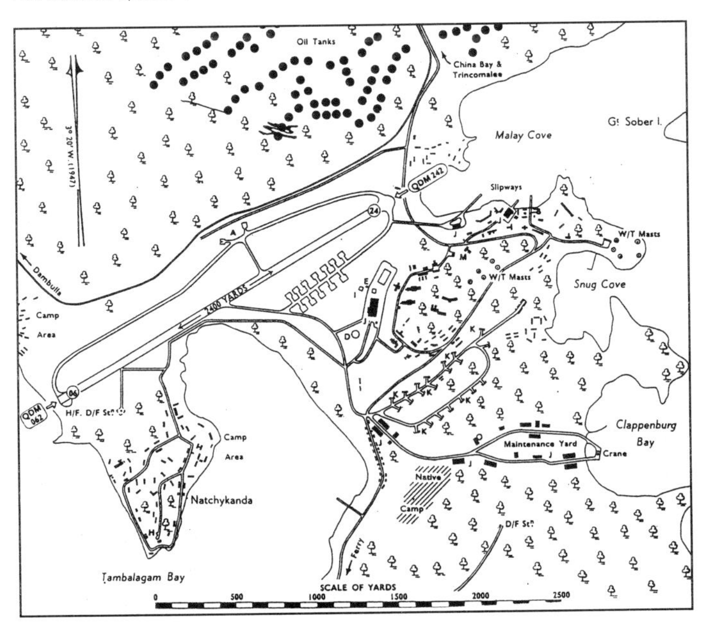

China Bay/Trincomalee.

CASTLE ARCHDALE/LOUGH ERNE

Opened in February 1941, within days this base had been renamed as RAF Lough Erne (although the name was changed back to Castle Archdale in January 1943). The first Sunderland unit, 201 Squadron, arrived in September 1941; various Sunderland units used the base during the war, the longest-serving resident being 423 Squadron.

CHINA BAY/TRINCOMALEE

Opened August 1939 as a landplane and seaplane base as part of the defences for Ceylon. The slipway was not added until June 1941, although 230 Squadron had used the alighting area from October 1939 to May 1940. Facilities were expanded in 1943 to cater for two flying boat squadrons and up to six landplane squadrons. From late 1949 the Sunderlands of 205 Squadron were using the base for detachments.

COLOMBO, CEYLON

DAR ES SALAAM

The main operational period was late 1943 to early 1945 in order to counter German submarine operations in the area of the Madagascar Channel.

FREETOWN

FURA BAY, SIERRA LEONE

GIBRALTAR

This location was in use from 1915 by Admiralty flying boats on anti-submarine patrols. It remained in use during the inter-war period as an RAF seaplane base, and from the late 1920s was also used by RAF flying boats. Sunderland detachments operated from mid-1940, 228 Squadron conducting patrols in the Western Mediterranean and the Atlantic. Other Sunderland units had detachments here at various times during the war.

Gibraltar.

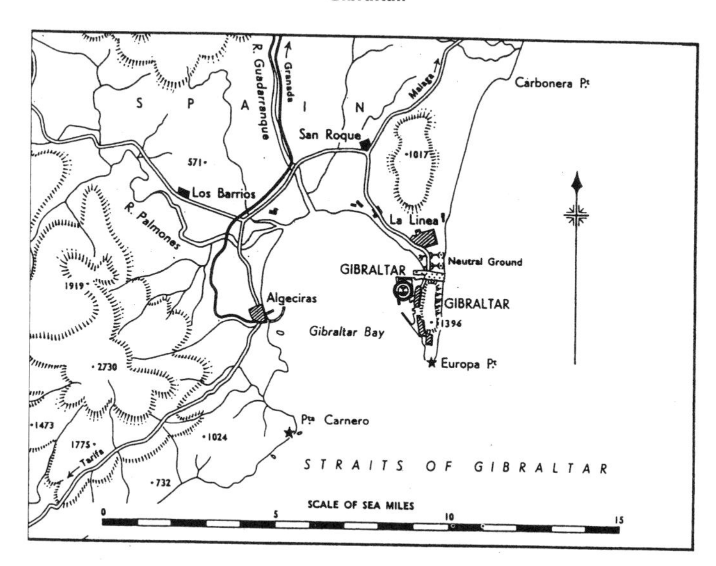

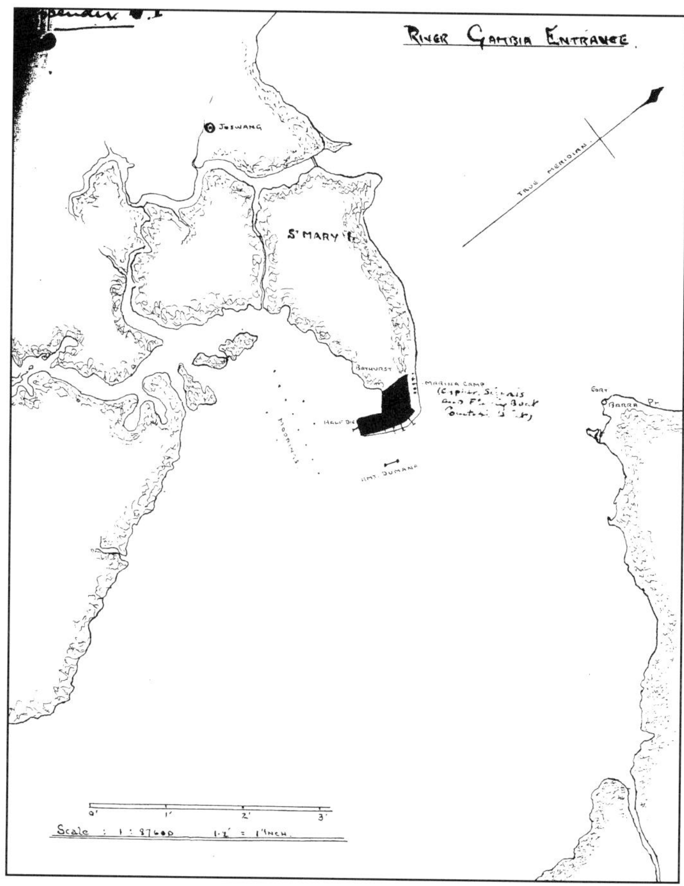

Half Die, Gambia.

HALF DIE, GAMBIA

HAMWORTHY

This Poole Harbour location was used throughout the war, primarily by civilian flying boats of BOAC. However, congestion at other bases meant that Hamworthy was used for diversions and detachments, plus, from late summer 1942, as home to 461 Squadron. The squadron moved to Pembroke Dock the following April and Hamworthy was used by Catalinas. In January 1944 the site was transferred to Transport Command, civil Sunderlands continuing to operate from here. BOAC ceased operations from here in March 1948.

INVERGORDON – *see* ALNESS

IWAKUNI, JAPAN

Used as a detachment base by the Far East Flying Boat Wing during the Korean War.

JUI

KAI TAK, HONG KONG

A site at Hong Kong was opened in 1927 with an airfield and moorings in Kowloon Harbour; it became home to a flight of Fairey IIIFs and, a few years later, flying boat detachments. No Sunderlands were based here before Hong Kong

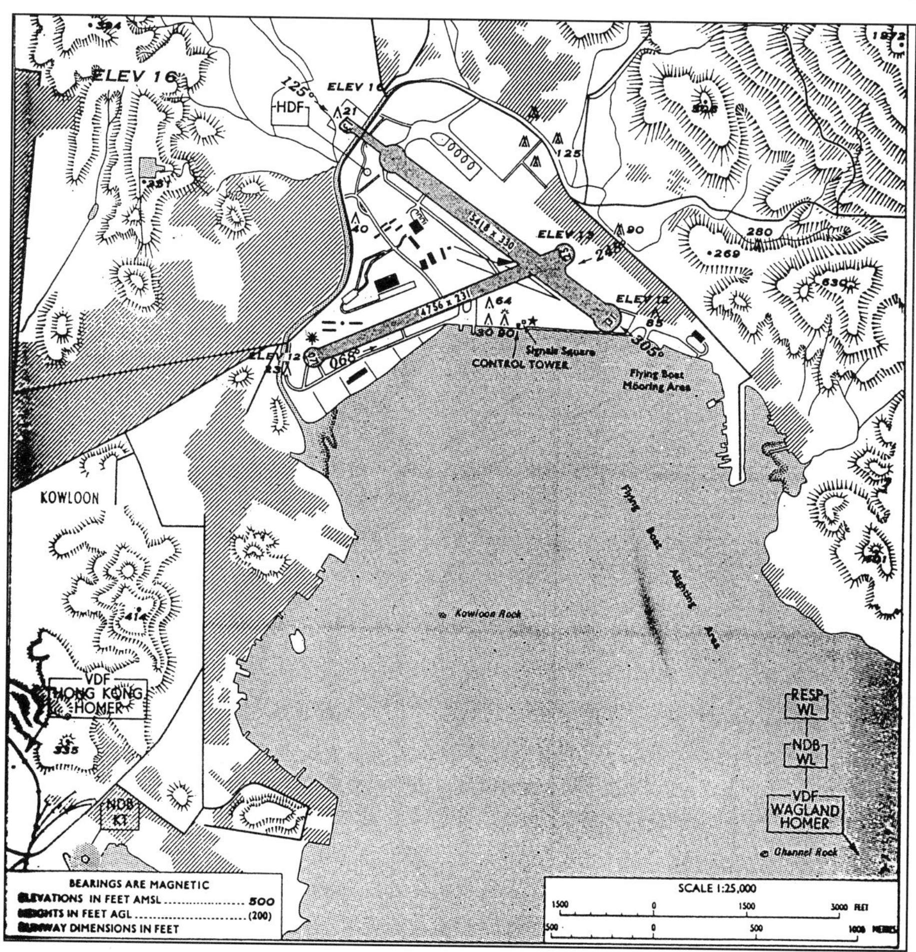

Kai Tak, Hong Kong.

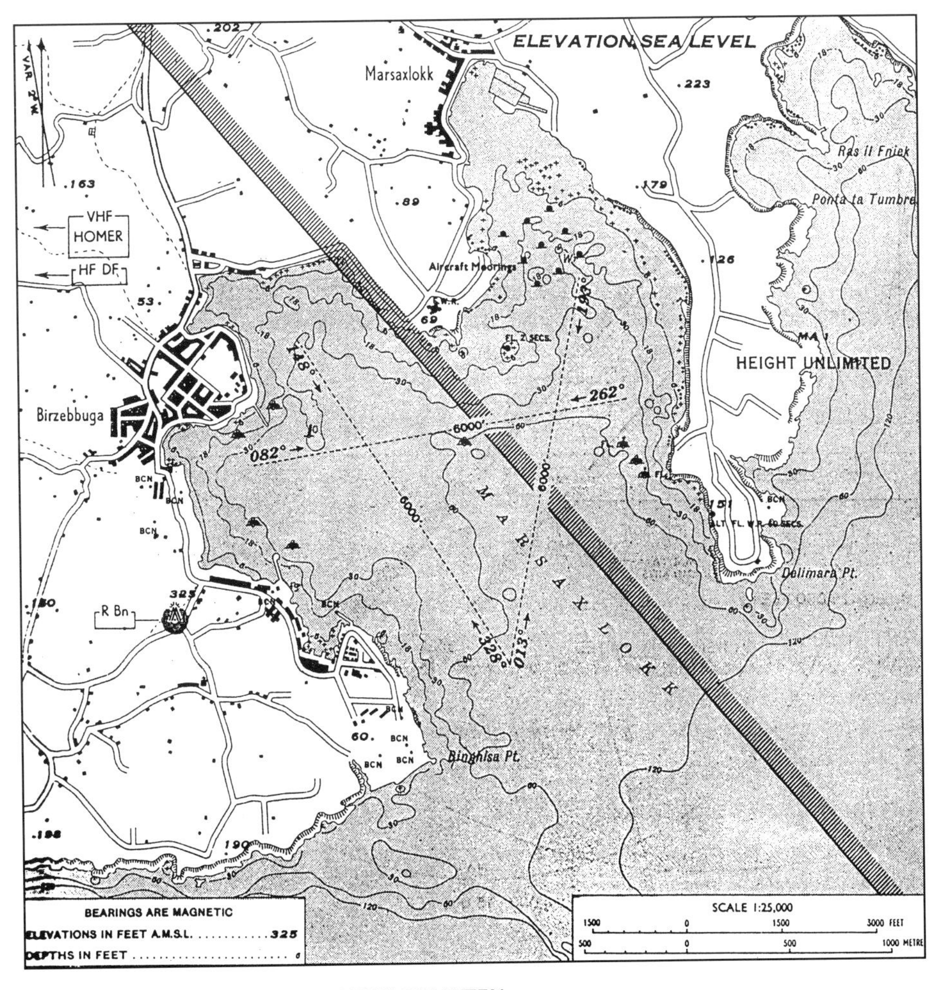

Marsaxlokk Bay, Kalafrana, Malta.

fell to the Japanese in December 1941. However, post-war Kai Tak became home to the Sunderland-equipped 1430 Flight, primarily for transport duties. This flight became 88 Squadron in September 1946, the squadron subsequently being involved with operational flying during the Malayan Emergency and the Korean War.

KALAFRANA, MALTA

As an Admiralty base this site opened in July 1916 in Marsa Scirocco Bay, Malta, initially for Curtiss H4 Small America flying boats. It subsequently transferred to RAF control from April 1918 as a seaplane base and remained operational in the inter-war period. First Sunderland use was by 228 Squadron from September 1940 and from that point on, although no Sunderland squadron was based there, it did house various detachments on an almost permanent basis up to late 1943. A number of other bays were also used as flying boat anchorages from time to time.

KASFAREET

KIPEVU

KOGGALA

One of the most important flying boat bases on Ceylon (Sri Lanka).

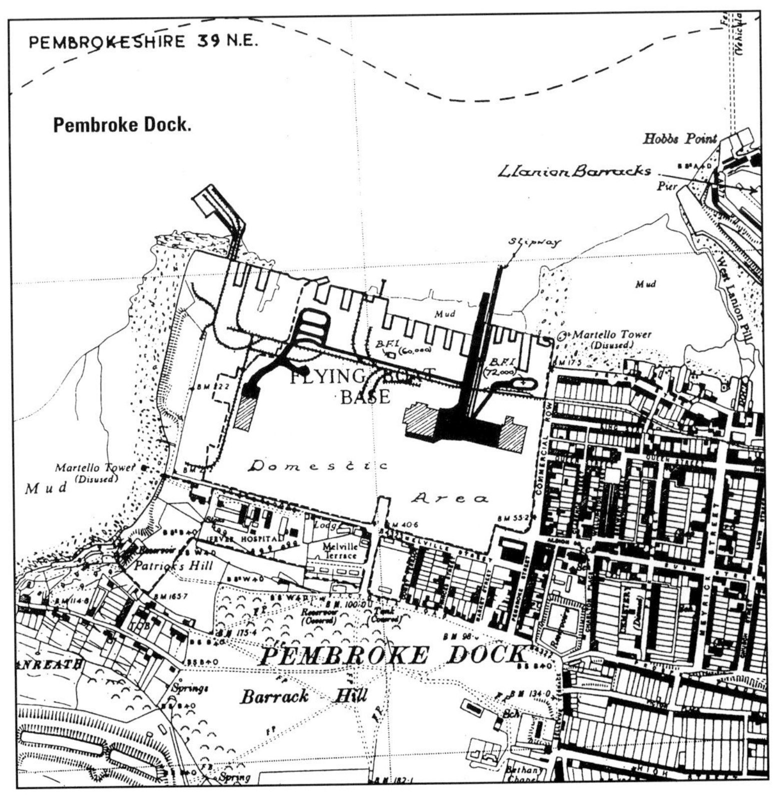

Pembroke Dock.

MOUNT BATTEN

The Cattewater was one of the best-known wartime bases for the Sunderland. Opened in World War One, the site remained active between the wars; the first Sunderland unit, 204 Squadron, arrived in June 1939. This squadron was replaced in April 1940 by 10 Squadron RAAF, this unit soon establishing a fine operational record. It was joined by a second Australian unit, 461 Squadron, in April 1942, although the latter soon moved to Hamworthy. With the end of the war, 10 Squadron returned to Australia (October 1945) and Mount Batten had no based flying boats, although other UK-based Sunderlands visited from time to time.

OBAN

In use from late 1937, the first Sunderland appeared at Oban in July 1940, and 210 Squadron was soon flying intensive operations. The following year saw 210 Squadron re-equip and move out, although 10 Squadron RAAF operated from Oban for a period. Aircraft of 228 Squadron arrived in March 1942 and stayed to the end of the year, being replaced by 330, 422 and 423 Squadrons at different times the following year. 1943 also saw the arrival of 302 FTU (amalgamating with 308 FTU the following year), this unit specializing in Sunderland training and remaining here until April 1945.

PEMBROKE DOCK

Opened in 1930, by the outbreak of war Pembroke Dock was a significant flying boat base, housing two Sunderland squadrons, Nos 201 and 228. It remained an operational base throughout the war, playing a full role in Coastal Command's war against the U-boats. Post-war it became the main UK Sunderland base, housing 201 and 230 Squadrons until they disbanded in March 1957 – at which point Pembroke Dock transferred to the Admiralty. However, the Short Sunderland Preservation Trust kept ML824 here from 1961 to 1976, at which time it moved to the RAF Museum at Hendon.

PENANG, MALAYSIA

PORT ETIENNE

Port Etienne was a 'miserable sandy hole with nothing much there except the refuelling facility and a radio beacon, and with the rather dodgy French attitude to the war we never got a friendly reception. BOAC arranged for a Shell barge to be stationed here.' (Howard Fry in *Best Years of Flying*)

RANGOON

REDHILLS LAKE

REYKJAVIK

SELETAR

Opened 1930 with land and sea facilities as RAF Base Singapore. Various units were stationed here during the 1930s, including the Singapore III flying

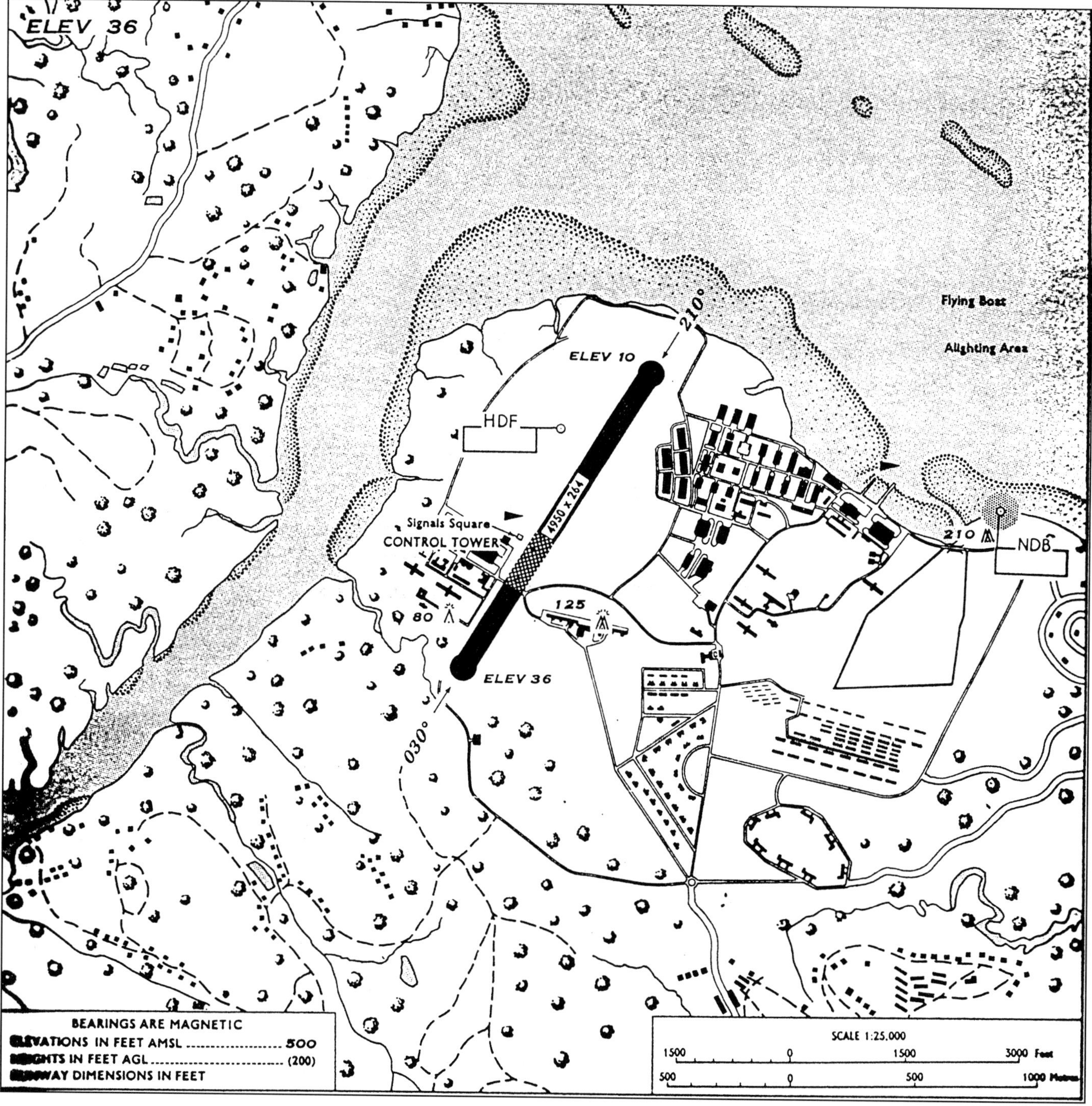

Seletar.

boats of 205 Squadron. The base had become RAF Seletar by the end of 1936. The first Sunderlands arrived in 1938 to re-equip 230 Squadron, starting a long association of this squadron with the Sunderland in the Far East. When 230 Squadron left for Ceylon it ended Sunderland presence at Singapore until after the defeat of the Japanese. Post-war, 205 and 209 Squadrons were based here and undertook operational sorties during the Malayan Emergency and Korean War. In June 1952, strength was boosted by the arrival of 88 Squadron and the creation of the Far East Flying Boat Wing.

SCARAMANGA

Located on Lake Eleusis near Athens, this mooring area was part of a Greek naval facility and was used by RAF Sunderlands in 1941 to mount operations over the Aegean Sea.

SULLOM VOE

This Shetland base was established in 1939 to provide air cover between Norway and Iceland, the first Sunderland unit being 201 Squadron in August 1939 (to November 1939), followed in spring 1940 by 210 Squadron, although only for a short time. Sunderland units came and went throughout the war, although in the latter months this was simply occasional visits by aircraft of 4 OTU. Reduced to Care & Maintenance in 1946, there was at least one, brief, Sunderland return to these waters in 1952 for an exercise.

WIG BAY

There was very little operational activity from this site, but it was used by a number of training units prior to the arrival of the FBSU in spring 1942 and 57 MU in October 1943. Hundreds of flying boats passed through Wig Bay for repair, storage, modification or disposal. At the end of the war the MU held over 150 flying boats, including a substantial number of Sunderlands, many of which ended up as scrap metal. The MU disbanded in October 1951, the work being taken over by Short Bros, although an RAF element remained to late 1947.

The author would be delighted to hear from anyone who can add further details regarding these locations – recollections, photographs or plans/maps.

APPENDIX IV

The U-boat War

The following U-boats are credited in the Coastal Command record as having been hit by Sunderlands.

DATE	U-BOAT	SQN	AIRCRAFT	PILOT	GROUP	RESULT
30/1/40	U-55	228	N9025	Fg Off E J Brooks	15 Gp	sunk
1/7/40	U-26	10 RAAF	P9603	Flt Lt W N Gibson	15 Gp	sunk
16/8/40	U-51	210	P9603	Fg Off E R Baker	15 Gp	damage
5/6/42	U-71	10 RAAF	W3986	Flt Lt S R C Wood	19 Gp	damage
7/6/42	*Luigi Torelli*	10 RAAF	W3994	Plt Off T A Egerton	19 Gp	damage
		10, RAAF	W4019	Flt Lt E St C Yeoman	19 Gp	damage
11/6/42	U-105	10 RAAF	W3993	Flt Lt E B Martin	19 Gp	damage
1/9/42	*Reginaldo Guiliano*	10 RAAF	W3986	Flt Lt S R C Wood	19 Gp	damage
		10 RAAF	W3983	Flt Lt H G Pockley	19 Gp	damage
14/9/42	*Alabastro*	202	W6002	Fg Off E P Walshe	Gib	sunk
1/5/43	U-415	461	DV968	Flt Lt E C Smith	19 Gp	damage*2
2/5/43	U-332	461	DV968	Flt Lt E C Smith	19 Gp	sunk
7/5/43	U-465	10 RAAF		Flt Lt Rossiter	19 Gp	sunk
12/5/43	U-456	423	W6006	Flt Lt J Musgrave	15 Gp	sunk *1
24/5/43	U-441	228	EJ139	Fg Off H J Debden	19 Gp	damage
31/5/43	U-563	228	DD838	Fg Off W M French	19 Gp	sunk *3
		10 RAAF	DV969	Flt Lt M S Mainprize	19 Gp	
31/5/43	U-440	201	DD835	Flt Lt D M Gall	15 Gp	sunk
13/6/43	U-564	228	DV967	Fg Off L B Lee	19 Gp	damage*4
27/6/43	U-518	201	W6005	Fg Off B E Layne	15 Gp	damage
13/7/43	U-607	228	JM708	Fg Off R D Hanbury	19 Gp	sunk
30/7/43	U-461	461	W6077	Fg Off D Marrows	19 Gp	sunk
1/8/43	U-454	10 AAF	W4020	Flt Lt K G Fry	19 Gp	sunk
1/8/43	U-383	228	JM678	Flt Lt S White	19 Gp	sunk
2/8/43	U-106	228	JM708	Fg Off R D Hanbury	19 Gp	sunk
		461	DV968	Flt Lt I A F Clarke	19 Gp	sunk
4/8/43	U-489	423	DD859	Fg Off A A Bishop	15 Gp	sunk
8/10/43	U-610	423	DD863	Fg Off A Menaul	15 Gp	sunk
17/10/43	U-470?	422	JM712	Flt Lt P T Sargent	15 Gp	damage
8/1/44	U-426	10 RAAF	EK586	Fg Off J P Roberts	19 Gp	sunk
28/1/44	U-571	461	EK577	Flt Lt R D Lucas	19 Gp	sunk
10/3/44	U-625	422	EK591	W/O W F Morton	15 Gp	sunk
24/4/44	U-672	423	DD862	Flt Lt F G Fellows	18 Gp	damage
16/5/44	U-240	330	JM667	Sqn Ldrt C T Johnsen	18 Gp	sunk
21/5/44	U-995	4(C) OTU	'S'	Plt Off E T King	18 Gp	damage
24/5/44	U-921	423	DW111	Flt Lt R H Nesbitt	18 Gp	damage
7/6/44	U-955	201	ML760	Flt Lt L H Baveystock	19 Gp	sunk
7/6/44	U-970	228	ML877	Flt Lt C G D Lancaster	19 Gp	sunk
11/6/44	U-323	228	ML880	Flt Lt M E Slaughter	19 Gp	damage
8/7/44	U-243	10 RAAF	W4030	Fg Off W B Tilley	19 Gp	sunk
11/7/44	U-1222	201	ML881	Flt Lt I F B Walters	19 Gp	sunk
19/7/44	U-387	330	EJ155	Lt B Thurmann-Nielson	18 Gp	damage
11/8/44	U-385	461	ML741	Plt Off I F Southall	19 Gp	sunk
12/8/44	U-270	461	ML735	Fg Off D A Little	19 Gp	sunk

18/8/44	U-107	201	EJ150	Flt Lt L H Baveystock	19 Gp	sunk
6/12/44	U-297	201	'Y'	Flt Lt Hatton		sunk
30/4/45	U-242	201	ML783	Flt Lt K H Foster	15 Gp	sunk

*1 - previously damaged by 86 Sqn Liberator

*2 - with 172 Sqn Wellington and 612 Sqn Whitley

*3 - with 59 Sqn Halifax

*4 - sunk next day by Whitley

Records for the other operational theatres are not as specific and thus credits are more difficult to determine. The following successes were credited to Sunderlands operating in the Mediterranean:

Date	*Target*	*Sqn*	*Aircraft*	*Captain*	*Result*
28/6/40	*Argonauta*	230	L5804	Flt Lt Campbell	sunk
29/6/40	*Rubino*	230	L5804	Flt Lt Campbell	sunk
30/9/40	*Gondar*	230	L2166	Flt Lt Alington	sunk
9/1/42	U-577	230	W3987	Sqn Ldr Garside	sunk
13/6/42	*Otario*	202	W4028		damage
14/9/42	*Alabastro*	202	W6002		sunk*

*also credited in Coastal Command record

THE U-BOATS

U-26
Type IA, 862 tons, Captain: Scheringer
Sunk 1 Jul 1940, N4803 W1130
First war cruise 29 Apr 40
Sunk on sixth cruise, left 20 Jun 40 for SW Ireland

U-55
Type VIIB, 753 tons, Captain: Heidel
Sunk 30 Jan 1940, N4837 W0746
Sunk on first war cruise, left 18 Jan 40

U-106
Type IXB, 1,051 tons, Captain: Dammerow
Sunk 2 Aug 1943, N4635 W1155
First war cruise 4 Jan 41 – 10 Feb 42
Sunk on tenth cruise, left Lorient 28 Jul 43 for Atlantic

U-107
Type IXB, 1,051 tons, Captain: Fritz
Sunk 18 Aug 1944, N4646 W0339
First war cruise 25 Jan 41 – 1 Mar 41
Sunk on thirteenth cruise, left Lorient 16 Aug 44 for La Pallice

U-240
Type VIIC, 769 tons, Captain: Link
Sunk 16 May 1944, N6305 E0310
Sunk en route Bergen to Narvik

U-242
Type VIIC, 769 tons, Captain: Riedel
Sunk 30 Apr 1945, N5342 W0453
First war cruise 8 Jun 44 – 26 Jun 44
Sunk on second Atlantic cruise, left Kristiansand 4 Mar 45 for Irish Sea

U-243
Type VIIC, 769 tons, Captain: Maerten
Sunk 8 Jul 1944, N4706 W0640
First war cruise 8 Jun 44 – 11 Jun 44
Sunk on second cruise, left Bergen 15 Jun 44 for English Channel

U-270
Type VIIC, 769 tons, Captain: Schreiber
Sunk 12 Aug 1944, N4619 W0256
First war cruise 23 Mar 43 – 14 May 43
Sunk on fifth cruise, left Lorient 10 Aug 44 for repairs at La Pallice

U-332
Type VIIC, 769 tons, Captain: Huttemann
Sunk 2 May 1943, N4448 W0858
First war cruise 30 Oct 41 – 16 Dec 41
Sunk on seventh cruise, left La Pallice 26 Apr 43 for N Atlantic

U-383
Type VIIC, 769 tons, Captain: Kremser
Sunk 1 Aug 1943, N4724 W1210
First war cruise 17 Oct 42 – 9 Dec 42
Sunk on fourth cruise, left Brest 29 Jul 43 for Atlantic

U-385
Type VIIC, 769 tons, Captain: Valentiner
Sunk 11 Aug 1944, N4616 W0245
First war cruise 5 Apr 44 – 4 Jun 44
Sunk on second cruise, left St Nazaire 9 Aug 44 for English Channel

U-426
Type VIIC, 769 tons, Captain: Reich
Sunk 8 Jan 1944, N4647 W1042
First war cruise 5 Oct 43 – 29 Nov 43
Sunk on second cruise, left Brest 3 Jan 44 for Atlantic

U-440
Type VIIC, 769 tons, Captain: Schwaff
Sunk 31 May 1943, N4538 W1304
First war cruise 1 Sep 42 – 21 Sep 42
Sunk on fifth cruise, left St Nazaire 26 May 43 for Atlantic

U-454
Type VIIC, 769 tons, Captain: Hacklaender
Sunk 1 Aug 1943, N4536 W1023
First war cruise 21 Feb 42 to Arctic flotilla
Sunk on fifth Atlantic cruise, left La Pallice 29 Jul 43 to Atlantic

U-456
Type VIIC, 769 tons, Captain: Teichert
Sunk 12 May 1943, N4837 W2239
First war cruise 31 Jan 42 to join Arctic fleet
Sunk on second Atlantic cruise, left Brest 24 Apr 43 for Azores

U-461
Type XIV, U-tanker, 1,688 tons
Sunk 30 Jul 1943, N4542 W1100
First war cruise 21 Jun 42 – 16 Aug 42
Sunk on sixth cruise, left Bordeaux 27 Jul 43 for Atlantic with U-462 and U-504

U-465
Type VIIC, 769 tons, Captain: Wolf
Sunk 7 May 1943
First war cruise 17 Nov 42 – 21 Dec 42
Sunk on fourth cruise, left St Nazaire 29 Apr 43 for Atlantic

U-489
Type XIV U-tanker, 1,688 tons, Captain: Schmandt
Sunk 4 Aug 1943, N6111 W1438
Sunk on first cruise, left Kiel 22 Jul 43 for Atlantic

U-563
Type VIIC, 769 tons, Captain: Borchardt
Sunk 31 May 1943, N4635 W1040
First war cruise 31 Jul 41 – 10 Sep 41
Sunk on 8th cruise, left Brest 29 May 43 for Atlantic

U-571
Type VIIC, 769 tons, Captain: Luessow
Sunk 28 Jan 1944, N5241 W1427
First war cruise 18 Oct 41 – 26 Nov 41
Sunk on ninth cruise, left La Pallice 8 Jan 44 for Atlantic

U-577
Type VIIC, 769 tons, CO: Schauenburg
Sunk 9 Jan 1942
First war cruise 17 Oct 41 – 26 Nov 41
Sunk in Mediterranean

U-607
Type VIIC, 769 tons, Captain: Jeschonek
Sunk 13 Jul 1943, N4502 W0914
First war cruise 9 Jul 42 – 16 Aug 42
Sunk on fifth cruise, left St Nazaire 10 Jul 43 to mine off Jamaica

U-610
Type VIIC, 769 tons, Captain: Von Freyberg
Sunk 8 Oct 1943, N5545 W2433
Fist war cruise 12 Sep 42 – 31 Oct 42
Sunk fifth cruise, left St Nazaire 12 Sep 43 to Atlantic

U-625
Type VIIC, 769 tons, Captain: Straub
Sunk 10 Mar 1944, N5235 W2019
First cruise 1 Oct 42 to Arctic flotilla
Sunk on 2nd Atlantic cruise, left Brest 29 Feb 44 for Atlantic

U-1222
Type IXC, 1,414 tons, Captain: Bielfeld
Sunk 11 Jul 1944, N4631 W0529
Sunk on first war cruise returning from Canadian seaboard

THE U-BOAT TYPES

At the outbreak of the war the Germans had fifty-seven U-boats available and in the first year of operations lost twenty-eight and commissioned twenty-eight; however, the number of available boats fell to twenty-seven for the start of the second year of operations.

U-boat numbers averaged fewer than 200 for the period late 1942 to early 1944 (*see* table), the majority of these, until later in the war, being the Type VIIC.

Number of Operational U-boats

Oct 1941	80
Jul 1942	140
Oct 1942	196
Apr 1943	240
Oct 1943	175
Apr 1944	166
Oct 1944	141

(source: German Naval History – the U-boat war)

Type VII U-boat
This 517-ton boat had a range of 8,700 miles and, powered by two 1,400hp MAN diesel engines, could operate at 16kt on the surface and 8kt underwater. The first boat, U-29, was completed in April 1936. It carried 12–14 torpedoes, fired from four forward and one aft tubes. Overall it was a good U-boat and was built in three main variants – A, B and C – with various modifications, the Type VIIC for example, was 760 tons and carried deck armament of one 88mm and two 20mm guns, although this was later modified.

PHASES OF THE U-BOAT WAR

The following 'eight phases of the U-boat war' are an abbreviated summary of details contained in the August 1945 'Anti-submarine Report'. Whilst the details relate to all aspects of the Allied campaign against the U-boats, the summary does provide a useful framework within which to consider the Sunderland's anti-submarine war.

Period I, Sep 1939 – Jun 1940
The main effort is directed against shipping proceeding unescorted in the South-Western Approaches, but U-boats are also sent as far south as Gibraltar. In April 1940 Germany overran Denmark and Norway and obtained more bases for U-boats.

Period II, Jun 1940 – Mar 1941
Italy enters the war. A few days later France collapses and the enemy begins his four-year occupation of the Biscay harbours. In July U-boat 'aces' begin night attacks on convoys in the North-Eastern Approaches and during the succeeding months cause us [i.e. the British] very heavy losses.

Period III, Mar 1941 – Jan 1942
The loss of his three best captains in March demonstrates to the enemy the effectiveness of our counter-measures in the North-Western Approaches. He therefore disposes his U-boats further out in the Atlantic and exploits a 'soft spot' off the West African coast, thereby severely stretching our defences. In the summer of 1941 we institute escorted convoys. In November 1941 German U-boats were sent to the Mediterranean. Italian U-boats are operating both in the Mediterranean and the Atlantic.

Period IV, Jan 1942 – Jul 1942
On 12 January 1942 a large force of U-boats begins to operate off the Atlantic seaboard of the USA and the coastal areas of Nova Scotia and Newfoundland. Very heavy losses are suffered. Japanese U-boats appear in the Indian Ocean and sink merchantmen as far west as the Mozambique Channel.

Period V, Aug 1942 – May 1943
The enemy, now disposing of a very large number of U-boats, makes a prodigious effort to sever the convoy routes that link Great Britain with North America. The battle is fought through the worst autumn and winter of the war and in March the enemy comes within sight of victory but about the 21st his efforts begin to falter. By increasing the endurance of aircraft we close the gap in air cover south of Greenland; after two fierce battles in May the enemy withdraws from the North Atlantic.

Period VI, May 1943 – Sep 1943
The Allies go on the offensive and inflict crushing losses, our air/sea operations in the Bay of Biscay coming to a sudden stop in August literally for lack of targets. The North Atlantic is wiped clear of sinkings. In the Madagascar area the enemy achieves his best results.

Period VII, Sep 1943 – May 1944
The enemy endeavours to resume the offensive in the North Atlantic, but his attempts are quickly smothered and he achieves little, while suffering heavy losses. In the Indian Ocean German U-boats operate with considerable success.

Period VIII, May 1944 – May 1945
On 16 May 1944 Coastal Command takes the first step in the operations that are to lead to the destruction of Germany and starts an offensive against U-boats in Norwegian waters. Despite the schnorkel, which largely neutralizes our immense air power, the U-boats are prevented from impeding our invasion operation and in August are driven from the Biscay ports. After a lull, the enemy returns to the UK coastal waters and makes an effort which increases with the months. By considerable exertions the enemy's operations are countered and a large number of boats destroyed.

APPENDIX V

Sunderland Order of Battle

The following details are based on RAF Form SD161 'Location of Royal Air Force Units'. These do not always include detachments, for example there is no mention of any detachments at Malta, and the dates of transfer of a Squadron sometimes reflect the plan rather than the actuality!

SEP 1939

No. 15 Gp:
204 Sqn Mount Batten
210 Sqn Pembroke Dock
Middle East:
228 Sqn Alexandria (about to return to UK)

MAR 1940

No. 15 Gp:
204 Sqn Mount Batten
10 RAAF Pembroke Dock
210 Sqn Pembroke Dock
228 Sqn Pembroke Dock
Middle East:
230 Sqn Alexandria (arrived May 1940)

FEB 1941

No. 15 Gp:
10 RAAF det Oban
210 Sqn Oban
FBTS
No. 18 Gp:
201 Sqn Sullom Voe
204 Sqn Sullon Voe
No. 19 Gp:
10 RAAF Mount Batten

SEP 1941

HQ Coastal Command:
95 Sqn Freetown
204 Sqn Bathurst
No. 17 Gp:
4 OTU Invergordon
No. 18 Gp:
201 Sqn Sullom Voe
No. 19 Gp:
10 RAAF Pembroke Dock
Middle East Command (MEC):
230 Sqn Aboukir
West Africa Command:
204 Sqn Bathurst
228 Sqn Bathurst

FEB 1942

No. 15 Gp:
201 Sqn Lough Erne
228 Sqn Stranraer
228 det Lough Erne
No. 17 Gp:
4 OTU Invergordon
No. 19 Gp:
10 RAAF Mount Batten
Middle East Command (MEC):
230 Sqn Aboukir
West Africa Command:
95 Sqn Fura Bay
204 Sqn Half Die

OCT 1942

HQ Coastal Command:
202 Sqn Gibraltar
No. 15 Gp:
201 Sqn Lough Erne
228 Sqn Oban
422 Sqn Lough Erne (forming)
423 Sqn Oban (forming)
No. 17 Gp:
4 OTU Invergordon
No. 19 Gp:
10 RAAF Mount Batten
119 Sqn Pembroke Dock
461 Sqn Hamworthy
FBMM Pembroke Dock
Middle East Command (MEC):
230 Sqn Aboukir
West Africa Command:
95 Sqn Jui
204 Sqn Half Die

FEB 1943

No. 15 Gp:
201 Sqn Castle Archdale
228 Sqn Castle Archdale
246 Sqn Bowmore
330 Sqn Bowmore/Oban
422 Sqn Bowmore/Oban
423 Sqn Castle Archdale
No. 17 Gp:
4 OTU Invergordon

No. 19 Gp:

10 RAAF	Mount Batten
119 Sqn	Pembroke Dock
461 Sqn	Hamworthy

Gibraltar:

202 Sqn	Gibraltar

Middle East Command (MEC):

230 Sqn	Aboukir

West Africa Command:

95 Sqn	Jui
204 Sqn	Half Die

OCT 1943

No. 15 Gp:

201 Sqn	Castle Archdale
422 Sqn det	Castle Archdale
422 Sqn	Bowmore
423 Sqn	Castle Archdale

No. 17 Gp:

4 OTU	Alness
308 FTU	Pembroke Dock

No. 18 Gp:

330 Sqn	Sullom Voe

No. 19 Gp:

10 RAAF	Mount Batten
228 Sqn	Pembroke Dock
461 Sqn	Pembroke Dock

Middle East Command (MEC):

230 Sqn	Argouti/Dar es Salaam

West Africa Command:

95 Sqn	Bathurst/Port Etienne
204 Sqn	Bathurst/Port Etienne

FEB 1944

No. 15 Gp:

201 Sqn	Castle Archdale
422 Sqn	Castle Archdale
423 Sqn	Castle Archdale

No. 17 Gp:

4 OTU	Alness
308 FTU	Pembroke Dock

No. 18 Gp:

330 Sqn	Sullom Voe

No. 19 Gp:

10 RAAF	Mount Batten
228 Sqn	Pembroke Dock
461 Sqn	Pembroke Dock

Middle East Command:

230 Sqn	Dar es Salaam

West Africa Command:

95 Sqn	Bathurst/Port Etienne
204 Sqn	Bathurst/Port Etienne
343 French Naval	Dakar

OCT 1944

No. 15 Gp:

422 Sqn	Castle Archdale
423 Sqn	Castle Archdale

No. 17 Gp:

4 OTU	Alness
231 OTU	Killadeas
302 FTU	Oban

No. 18 Gp:

330 Sqn	Sullom Voe

No. 19 Gp:

10 RAAF	Mount Batten
201 Sqn	Pembroke Dock
228 Sqn	Pembroke Dock
461 Sqn	Pembroke Dock

[All squadrons except 422 and 423 were designated MR, range up to 1,400nm, the other two were LR, range 1,400nm–2,000nm.]

Middle East Command (AHQ West Africa):

95 Sqn	Bathurst/Port Etienne
204 Sqn	Jui/Fishlake/Point Noire/Abidjan
270 Sqn	Abidjan/Point Noire/Libreville
343 French Naval	Dakar/Port Etienne
490 Sqn	Jui/Abidjan/Fishlake

Air Command South East Asia (ACSEA):

230 Sqn	Koggala

FEB 1945

No. 15 Gp:

201 Sqn	Castle Archdale
423 Sqn	Castle Archdale

No. 17 Gp:

4 OTU	Alness
131 OTU	Killadeas
302 FTU	Oban

No. 18 Gp:

330 Sqn	Sullom Voe

No. 19 Gp:

10 RAAF	Mount Batten
228 Sqn	Pembroke Dock
422 Sqn	Pembroke Dock
461 Sqn	Pembroke Dock

Middle East Command (AHQ West Africa):

95 Sqn	Bathurst/Port Etienne
204 Sqn	Jui/Abidjan
270 Sqn	Libreville/Abidjan/Fishlake
343 French Naval	Dakar/Port Etienne
490 Sqn	Jui/Abidjan/Fishlake

Air Command South East Asia (ACSEA):

230 Sqn	Koggala

FBTS – Flying Boat Training Squadron
FBMM – Flying Boat Major Maintenance

Operational Strength of Sunderlands in Coastal Command

The following data is from Coastal Command return Form 1723.

Jan 41	13 serviceable aircraft
Jul 41	14 serviceable aircraft
Dec 41	17 serviceable aircraft
Jul 42	24 serviceable aircraft
Dec 42	19 serviceable aircraft
Jul 43	27 serviceable aircraft (63 operational establishment)
Dec 43	64 serviceable aircraft (63 OE)
Jul 44	101 serviceable aircraft (81 OE)
Dec 44	87 serviceable aircraft (81 OE)

DATE	SORTIES	HOURS	LOSSES
1940			
May	43		
Jun	110		
Jul	109		
Aug	78		
Sep	80		
Oct	66		
Nov	43		
Dec	46		
TOTAL	*575*		
1941			
Jan	106	811	
Feb	98	832	
Mar	178	1543	
Apr	145	1254	
May	134	1154	
Jun	146	1256	
Jul	108	925	
Aug	109	874	1
Sep	82	729	
Oct	97	736	
Nov	47	384	
Dec	58	539	
TOTAL	*1308*	*11,037*	*1*
1942			
Jan	27	269	
Feb	61	595	1
Mar	81	850	
Apr	123	1323	1
May	100	1036	
Jun	91	936	1
Jul	161	1800	1
Aug	159	1695	2
Sep	143	1623	1
Oct	120	1259	
Nov	209	2366	
Dec	193	1972	1
TOTAL	*1468*	*15,724*	*8*

DATE	SORTIES	HOURS	LOSSES
(no details of hours flown given for subsequent years)			
1943			
Jan	186		1
Feb	161		
Mar	251		
Apr	239		
May	306		4
Jun	312		3
Jul	363		2
Aug	299		6
Sep	324		3
Oct	270		2
Nov	280		5
Dec	275		1
TOTAL	*3266*		*27*
1944			
Jan	230		1
Feb	262		
Mar	349		2
Apr	277		
May	291		1
Jun	579		3
Jul	458		
Aug	525		
Sep	501		
Oct	416		2
Nov	318		
Dec	303		
TOTAL	*4509*		*9*
1945			
Jan	280		
Feb	229		
Mar	382		
Apr	396		
May	115		1
TOTAL	*1402*		*1*

APPENDIX VII

Coastal Command Manual of Anti-U-boat Warfare

The May 1944 edition of this manual contained tactical notes to be employed by Coastal Command crews in the war against the U-boats, all were relevant to Sunderland crews and these extracts give a flavour of the operational content of the manual.

No. 4. Sighting and Approach
Day. The first aim of the pilot on sighting a suspicious object or obtaining a likely radar contact must be to get in his attack as quickly as possible. Subject to this consideration, his second aim must be to remain undetected for as long as possible during his approach. The line of approach to the target from the point of sighting must always be direct except on the rare occasions on which the target is sighted at such close range as to make it impossible for the pilot to lose sufficient height in a direct approach. On such occasions it is permissible to consider the advantage of hammering the U-boat's gunners by making an approach down sun or down wind, if the wind is strong and the sea rough.

U-boat manoeuvres. If a U-boat commander does not intend, or is unable, to submerge on sighting hostile aircraft, his first consideration – in order to give his gunners the maximum field of fire – is to manoeuvre his ship in such a way as always to keep the aircraft abaft his beam. If he keeps his head, he will nearly always succeed in doing this, since the U-boat's speed of turning is much greater than that of any aircraft in proportion to their relative turning circles. In practice it has been found that the aircraft is almost invariably forced to attack from some direction between the beam and stern.

Machine-gun fire from aircraft. Fire should be brought to bear on the U-boat's gun crews from the maximum range – i.e. about 1,200yd. Experience has shown that fire from .303 and .5 guns is fully effective at 1,000yd. Machine-gun fire from aircraft, especially from the front guns during the actual run-in, has in the past proved so effective that whole gun crews have on occasions been killed or wounded and flak has ceased by the time the aircraft released its depth charges.

Evasive action. When attacking with depth charges it is possible for the aircraft to take evasive action against the U-boat's flak while making its approach. It is emphasized that the only effective form of evasive action is 'undulating' – alternately losing and gaining 2–300ft in height while maintaining a steady course. This interferes least with the bombing run and with the aircraft's gunners. It provides the easiest method of avoiding a fixed or moving box barrage. All evasive tactics must be deliberate and well-defined.

Should the aircraft be very badly damaged by flak on the run-in, it may become necessary to break off the attack, but it is worth noting that, unless the damage appears serious, it is generally more dangerous to break off an attack than to press it home.

No. 5. Depth Charge and A/S Bomb attacks
Judgement of Approach. The key to a successful attack is the right judgement of the approach. It is essential that in the latter stages of the attack the aircraft should be tracking towards the target at the right height and travelling at as nearly as possible the speed set on the distributor. Bomb doors must be open, switches checked and the aircraft and crew must be ready in all aspects for the attack. When ¾ mile to a mile from the target the bomb-aimer should already be in position. In the event of a sighting at very close range, however, neither approach nor attack should ever be delayed on this account; it is better for the pilot to bomb by eye than lose time waiting for the bomb-aimer.

Point of aim – surfaced U-boats. In all attacks the aircraft should be steered to track over the conning-tower. Experience has shown that when the pilot attempts to make allowance for the forward movement of the U-boat by tracking ahead of the conning tower, the most frequent result is a miss just ahead of the U-boat. If you track directly over it you will kill the U-boat.

Point of aim – submerged U-boats. The advantage of an attack up or down track on a submerged U-boat are considerable. When attacking from a low level and without a bomb sight, all calculations as to the position of the U-boat can be ignored and, provided that a stick of at least 300ft is used, the swirl provides the aiming point – for the first depth charge in the case of an up-track attack from astern and for the last in the case of a down-track attack from ahead. When carrying out beam, bow or quarter attacks it becomes necessary to calculate the probable location of the U-boat more exactly. Taking the 100ft long swirl left by the conning-tower as a yardstick, all that is necessary for the pilot to do when he sees that the U-boat has dived, is to estimate whether he will cross track within 10 seconds of its submersion. If so he will aim to cross it one swirl's length ahead of the swirl. If more than 10 seconds, he should aim to drop his depth charges two swirl lengths ahead.

No. 7. Depth Charges and A/S bombs – characteristics
The Mk IX Torpex Depth Charge is in effect such a flexible weapon that it can, for all practical purposes, be dropped from any height up to 1,000ft and at any speed likely to be reached by the normal type of Anti U-boat aircraft. A stick of six Mk XI Torpex DCs spaced at 36ft and released from 50ft at 120kt yield the following results:

1. Height of impact splashes is about 60ft.
2. Detonation takes place 3 seconds after impact.
3. Dome forms 4 seconds after impact and is 210ft in diameter.

4. Dome breaks into plume 4.5 seconds after impact.
5. Plume attains full height 6 seconds after impact and is 2–300ft high.
6. Plume falls off 12 seconds after impact and has subsided 20 seconds after impact.
7. The disturbance of the surface of the sea persists for about 3 minutes in good sea conditions. After all traces of it have disappeared, the DC residue – a brownish scum – may remain for a considerable time on a calm sea.

Preserved Aircraft

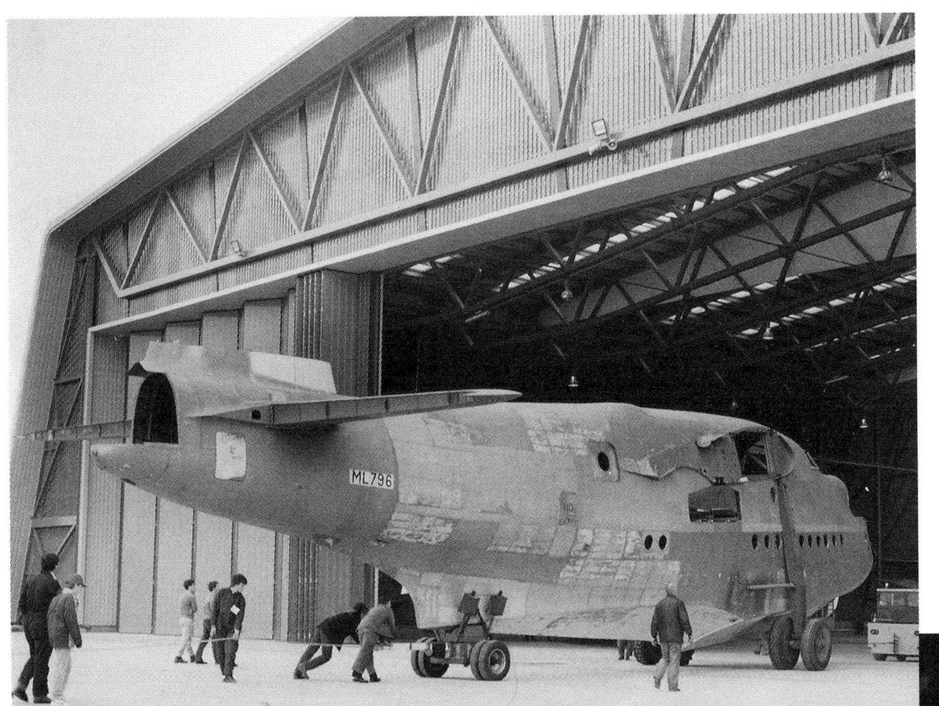

The sole airworthy Sunderland:
ML814 (NZ4108) – Fantasy of Flight, Polk City, Florida.

Preserved in museums:
ML796 – Imperial War Museum, Duxford, UK.

ML824 – RAF Museum, Hendon, UK.

JM715 – Hall of Aviation, Southampton, UK. Converted to Sandringham 4.

JM719 – Musée de L'Air, France. A Sunderland III that became a Sandringham and saw service as G-AKCO, VH-APG and F-OBIP.

Two views of ML796 entering the hangar at Duxford ready to undergo restoration. FlyPast Collection

NZ4115 (SZ584) preserved in New Zealand; further restoration work has been conducted on this aircraft since this photograph was taken. FlyPast Collection

Sandringham 4 VH-BRC (ex-JM715) in the Hall of Aviation, Southampton. FlyPast Collection

SZ584 (NZ4115) – MoTat. Having been built by Short Bros as a Mk V the aircraft was allocated to BOAC and flew with them as G-AHJR before being returned to the RAF for storage. It subsequently went to the RNZAF as NZ4115 in September 1953 and served with 5 Squadron before eventually being donated to MoTat in 1966. Significant restoration work has been undertaken in the 1990s.

VB880 (NZ4111) – RNZAF Museum.

VB881 (NZ4112) – Ferrymead Aeronautical Society (cockpit area only).

France – A number of Sunderlands have been recorded in various states of preservation (deterioration in some cases) in France but current information has proved difficult to obtain – the author would be delighted to receive any updates.

The RAF Museum houses Sunderland V ML824. FlyPast Collection

The sole airworthy Sunderland, ML814, is now resident at Fantasy of Flight, Florida. FlyPast Collection

Index

The names of ships and submarines are shown in *italics*. Not all references to inclusions in the appendices are listed here; for example the squadron details given in Appendix V.